ENZYME CHEMISTRY OF PHENOLIC COMPOUNDS

ENZYME CHEMISTRY OF PHENOLIC COMPOUNDS

Proceedings of the
Plant Phenolics Group Symposium
Liverpool, April 1962

Edited by

J. B. PRIDHAM

Lecturer in Organic Chemistry,
Royal Holloway College,
University of London

A Pergamon Press Book

THE MACMILLAN COMPANY
NEW YORK
1963

THE MACMILLAN COMPANY
60 Fifth Avenue,
New York 11, N.Y.

This book is distributed by
THE MACMILLAN COMPANY · NEW YORK
pursuant to a special arrangement with
PERGAMON PRESS LIMITED
Oxford, England

Library of Congress Card No. 63-12795

Printed in Great Britain by J. W. Arrowsmith Ltd., Bristol, England

PREFACE

THE interaction of enzymes and phenolic compounds was discussed at a Symposium of the Plant Phenolics Group which was held at Rathbone Hall, University of Liverpool, during 11–12 April 1962. The majority of papers presented were concerned either directly or indirectly with oxidation-reduction systems in plants and microorganisms, and particular emphasis was laid on the phenolase complex which is now being studied extensively by investigators of various disciplines using a variety of modern techniques.

There is still much confusion over the nomenclature used for these enzymes and the reactions which they catalyse. It was apparent from the discussions which followed the papers, however, that there is a renewed and widespread interest in this topic and it is to be hoped that many of the present difficulties will soon be overcome.

As a result of the sudden and tragic death of Dr. E. A. H. Roberts, Chairman of the Group, the expected paper on Tea Oxidases was not presented. Instead, a contribution from Professor J. Lavollay and his colleagues on Potato Polyphenol Oxidases was read by Dr. T. Swain.

We are indebted to the Chemical Society and Dr. E. R. Redfearn for permission to publish Fig. 6, page 61, and to Academic Press and Professor K. Wallenfels for allowing us to reproduce Figs. 3 and 4, page 78. Finally, I should like to thank my colleagues in the Chemistry and Botany Departments of Royal Holloway College for much helpful advice during the preparation of this book.

Royal Holloway College, 1962 J. B. PRIDHAM

CONTENTS

LIST OF CONTRIBUTORS

D. S. BENDALL, Department of Biochemistry, The University, Cambridge, England.

SHEILA M. BOCKS, Dyson Perrins Laboratory, The University, Oxford, England.

J. G. BOSWELL, Department of Botany, The University, Sheffield, England.

B. R. BROWN, Dyson Perrins Laboratory, The University, Oxford, England.

N. A. BURGES, Botany Department, The University, Liverpool, England.

J. GREEN, Walton Oaks Experimental Station (Vitamins Ltd.), Dorking Road, Tadworth, Surrey, England.

R. P. F. GREGORY, Department of Biochemistry, The University, Cambridge, England.

F. W. HEMMING, Department of Biochemistry, The University, Liverpool, England.

A. C. HULME, Agricultural Research Council, Ditton Laboratory, Larkfield, Maidstone, Kent, England.

H. M. HURST, Department of Botany, The University, Liverpool, England.

J. B. JEPSON, Courtauld Institute of Biochemistry, The Middlesex Hospital, Medical School, London, England.

J. D. JONES, Agricultural Research Council, Ditton Laboratory, Larkfield, Maidstone, Kent, England.

E. KÜSTER, Department of Industrial Microbiology, University College, Dublin, Eire.

J. LAVOLLAY, Conservatoire National des Arts et Métiers, 292, Rue Saint-Martin, Paris, France.

G. LEGRAND, Conservatoire National des Arts et Métiers, 292, Rue Saint-Martin, Paris, France.

G. LEHONGRE, Conservatoire National des Arts et Métiers, 292, Rue Saint-Martin, Paris, France.

R. A. MORTON, Department of Biochemistry, The University, Liverpool, England.

J. NEUMANN, Conservatoire National des Arts et Métiers, 292, Rue Saint-Martin, Paris, France.

J. B. PRIDHAM, Department of Chemistry, Royal Holloway College, University of London, Englefield Green, Surrey, England.

A. H. WILLIAMS, Long Ashton Research Station, University of Bristol, Long Ashton, Bristol, England.

INTRODUCTORY ADDRESS

ENZYMES ASSOCIATED WITH PHENOLS

N. A. BURGES

The Hartley Botanical Laboratories, University of Liverpool

PHENOLIC substances in plants appear to be derived by at least three different pathways: via prephenic and shikimic acids, via mevalonic acid, and via an acetate condensation. Although the acetate hypothesis seems well substantiated[1] we still know nothing of the steps between the acetate stage and the first recognizable molecule which may contain 3–9 acetate units. One is tempted to take protein synthesis as an analogy and picture the acetate units being aligned on a jig, the linkages made more or less simultaneously and the completed molecule dislodged from the jig, enolization then giving rise to the phenolic starting points. Subsequently, simple steps lead to more complex molecules. Many of the enzymes responsible for these later steps and for those associated with the mevalonic and shikimic pathways seem to be fairly specialized and although some have been studied they have not attracted a great deal of interest. Similarly little has been published on the enzymes responsible for the degradative paths associated with the disappearance of complex phenolic derivatives. Most work has been concerned with the action and nature of oxidative enzymes on relatively simple phenolic materials and with their biological significance.

Interest centred first on the laccase enzyme from the Lack tree, *Rhus vernicifera*, and tyrosinase from fungi, towards the end of the last century. Since then numerous workers have investigated enzymes belonging to the general phenol oxidase group from numerous sources. Many of the studies were aimed at finding differential substrates which would enable the separation and classification of the enzymes involved. Among the substrates used *p*-cresol, catechol, quinol and tyrosine have received particular attention. It was soon realized, however, that other oxidases would also attack these substrates and, more recently, that characteristic phenol oxidases would attack non-phenolic molecules. The general nature and distribution of phenol oxidases have been well reviewed (e.g. ref. 2). However, several recent papers have appeared which help considerably in clarifying some points. Problems of nomenclature are also confusing and it is important that we adopt some degree of uniformity in our discussions.

It is common now to divide the phenol oxidases into two types; the tyrosinase type and the laccase type.

The typical tyrosinase acts on tyrosine (I) to give first 3,4-dihydroxyphenylalanine (II), then on further oxidation a quinone (III) which forms a heterocyclic red compound (IV) which is further oxidatively polymerized to melanin:

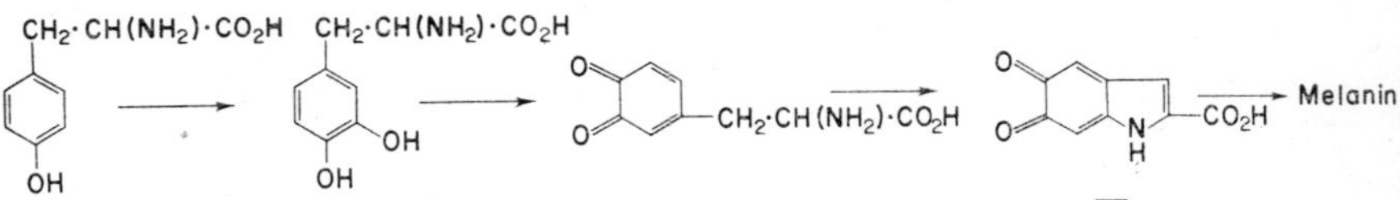

I II III IV

Comparable oxidations occur with other substrates and the action on *p*-cresol has been studied in considerable detail. Ability to carry out reactions of the type shown below as A has been termed cresolase activity and the B type as catecholase activity:

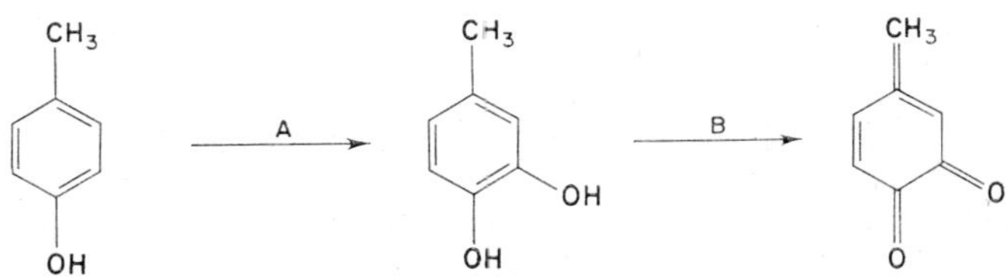

Although many investigators have attempted to isolate these as separate enzymes the general conclusion seems to be that the enzyme has two types of active centres and of these the cresolase centres are more easily inactivated.

Laccase was isolated first from the latex of the Japanese Lack tree. The latex underwent an enzymatic oxidative polymerization to form the black laquer of Japan. The major natural substrate in this reaction is urushiol (V). Similar laquers come from Indochina Lack where the phenol is laccol (VI) and Burma Lack which contains thitsiol (VII).

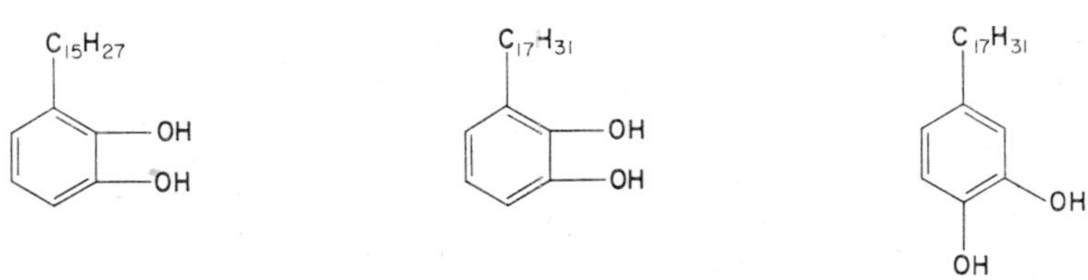

V VI VII

Subsequent studies have shown that enzymes of these types are widely distributed both in higher plants and fungi. Convenient summaries are given by Franke.[2] In the fungi Boiden[3] found laccase in 102 species and tyrosinase in 55; Lyr[4] examined 182 species and found 75 to be laccase-forming and 62 produced tyrosinase.

As knowledge accumulated it seemed that the two groups could be separated on the basis of their activity on certain substrates as shown in Table 1.

TABLE 1. CONVENTIONAL DIFFERENTIATION OF TYROSINEASE AND LACCASE

	Tyrosine	*p*-cresol	*p*-phenylene diamine	Quinol
Tyrosinase	+	+	—	—
Laccase	—	—	+	+

Although the above has been widely used to distinguish between laccase and tyrosinase it is clearly not entirely satisfactory. Various workers have, for instance, reported an action of typical laccase on *p*-cresol.[5]

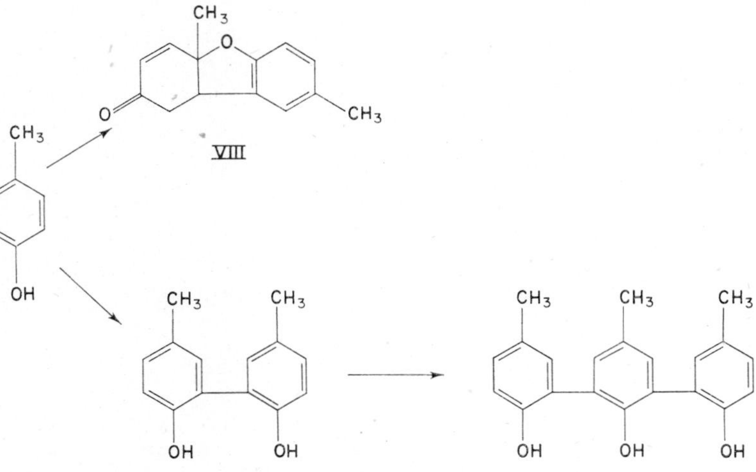

It has been known for some time that peroxidases in the presence of hydrogen peroxide will oxidize a wide range of mono- and di-hydric phenols. The action of peroxidase on *p*-cresol was examined by Westerfeld and Low[6] who concluded that three substances were formed although (VIII) was the main product.

Barton *et al.*[7,8] however, considered that the more probable formula for (VIII) was (IX).

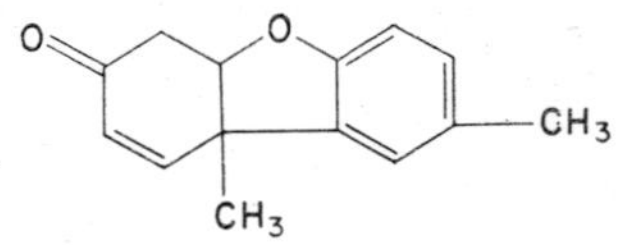

IX

Fåhraeus and Ljunggren[5] concluded that the white substance formed by the action of laccase on *p*-cresol was similar to that formed by peroxidase.

In many of the studies on tyrosinase it has been found that small quantities of dihydric phenols influence the activity. The work of Fåhraeus and Ljunggren[5] showed that *p*-cresol was oxidized very incompletely in a system containing substrate, buffer and the pure enzyme. If catechol were added either before the *p*-cresol or with it, then both substances were oxidized at a rapid rate. They concluded that the catechol was exercising a protective action on the cresolase activity and showed that this protective action could also be carried out with gelatin or Tween-80.

It would seem that both the tyrosinase and the laccase each have centres with cresolase and catecholase activity. In the case of laccase the more "natural" substrates like guaiacol and vanillin do not inactivate whereas more "artificial" substances like *p*-cresol and naphthol are strongly inactivating. On the basis of Fåhraeus and Ljunggren's work one may redefine the difference between tyrosinase and laccase in terms of their action on *p*-cresol.

Tyrosinase (cresolase activity):

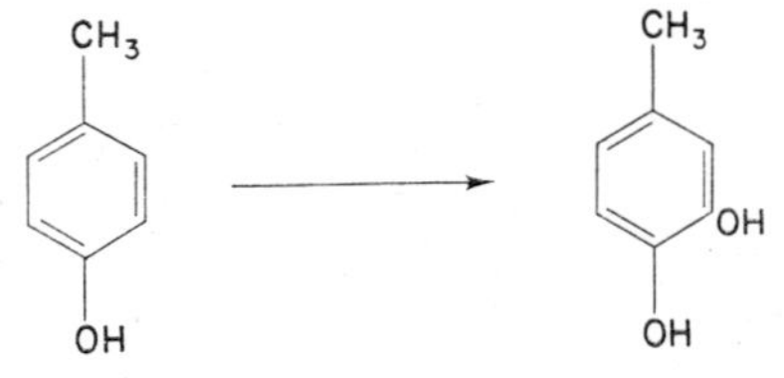

Laccase (cresolase activity):

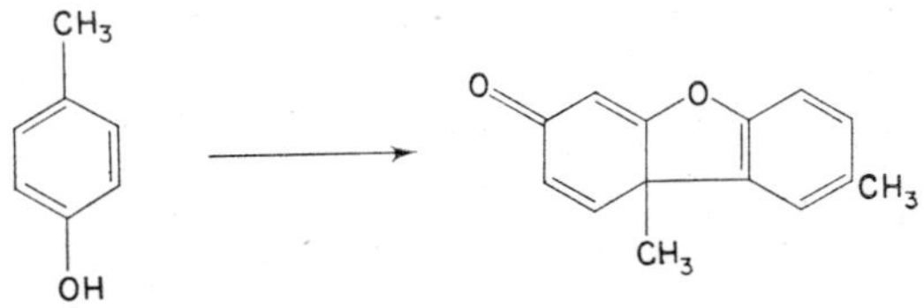

While both tyrosinase and laccase can attack *o*-dihydric phenols laccase also oxidizes *p*-dihydric phenols and trihydric phenols such as pyrogallol.[5]

Neufeld *et al.*[9] as the result of their work with crude extracts from *Pericularia oryzae* and *Polystictus versicolor*, concluded that not only an *o*-polyphenol oxidase was present but also a distinct *m*-polyphenol oxidase. The evidence for such an enzyme, however, seems very slight.

Our knowledge of the function of these enzymes is still unsatisfactory. Tyrosinase has from time to time been regarded as a terminal oxidase in some tissues, but the evidence seems conflicting. In *Polystictus* it does not seem to be associated with the respiratory processes.[10] In general polyphenol oxidases in higher plants become apparent only when the tissue is damaged. The typical reaction is the browning of bruised apple tissue and the blackening of peeled potatoes. In studies of laccase production by fungi, maximum enzyme activity has often been found to be associated with autolysis. Fåhraeus *et al.*[11] in their examination of laccase production were able to increase the enzyme level by factors of 50 or more, by the addition of inducers such as the xylidines, 2,5-xylidine being the most efficient inducer.

Evidence is accumulating to show that the lysozyme enzymes are associated with specific structures within the cell and are released only on breakdown of the lysosome. It may well be that there is a similar situation with the phenol oxidases. There is evidence in some plants that the enzyme and its substrates are separated perhaps into separate cells.[12]

In the case of the laccase enzyme in fungi it has been known for some time that there is a strong correlation between the ability to secrete laccase and to decompose lignin. Although this was pointed out by Bavendam over thirty years ago and has been amply confirmed by subsequent workers there is still no real evidence for the laccase being implicated in the lignin breakdown. The most commonly held belief seems to be that the laccase detoxicates the phenolic units produced in the initial rupture of the lignin molecule. In work now in progress on humic acid decomposition a similar correlation has been found between the decomposition of humic acid and the ability to reduce an aromatic carboxylic group.[13] All the humic acid decomposing fungi so far tested are laccase producers but the conditions under which they actively decolorize the humic acid are not those which seem to favour laccase production. It is still possible, however, that once the initial breakdown of the molecule has occurred laccase action may be induced and the enzyme detoxicates the phenolic moities produced.

In laboratory tests the phenolases are clearly capable of oxidative synthetic activity. In living organisms tyrosinases certainly seem to have a function in pigment formation in animals but the comparable reactions in plants seem less useful. Perhaps the ability to hydroxylate a monophenol may be important.

Mention has already been made of the action of peroxidases on *p*-cresol. These enzymes will oxidize quite a wide range of both mono- and di-hydric phenols. Seigel[14] showed that peroxidases were localized in the same regions as those in which lignin appeared to be formed. Compounds such as eugenol and thymol were oxidized to products with the characteristics of lignin. Catalase has also been shown to have the ability to oxidize phenols such as pyrogallol, *p*-cresol and catechol.

The tannases produced by fungi have been known for almost a century. The most abundant sources have been *Aspergillus niger* or species of *Penicillium*. The tannases are essentially hydrolytic esterases and the fungi which produce them remove and utilize the sugar element of the tannin leaving the gallic acid. Unless adequate additional sugar is supplied part or all of the gallic acid is also utilized. The general problem of hydrolytic enzymes will be discussed in detail by Dr. Pridham.

REFERENCES

1. R. W. RICKARDS, in *Recent Developments in the Chemistry of Natural Phenolic Compounds*, p. 1, ed. W. D. Ollis, Pergamon Press, London (1961).
2. W. FRANKE, *Encyclopedia of Plant Physiology*, Vol. 12, Pt. I, p. 401, Springer Verlag, Berlin (1960).
3. J. BOIDEN, *Rev. Mycol.* **16,** 173 (1951).
4. M. LYR, *Naturwiss.* **44,** 235 (1957).
5. G. FÅHRAEUS and H. LJUNGGREN, *Biochim. Biophys. Acta* **46,** 22 (1961) and **54,** 192 (1961).
6. W. W. WESTERFELD and C. LOWE, *J. Biol. Chem.* **145,** 463 (1942).
7. D. H. R. BARTON, A. M. DEFLORIN and O. E. EDWARDS, *J. Chem. Soc.* 530 (1956).
8. D. H. R. BARTON and T. COHEN, *Festschr. Arthur Stoll*, p. 117 (1959).
9. H. A. NEUFELD, F. M. LATTERFELL, L. F. GREEN and R. L. WEINTRAUB, *Arch. Biochem. Biophys.* **76**, 317 (1958).
10. D. BOULTER and A. BURGES, *Experientia* **11,** 188 (1955).
11. G. FÅHRAEUS, V. TULLANDER and H. LJUNGGREN, *Physiologia Plantarum* **11,** 631 (1958).
12. E. A. H. ROBERTS, *Adv. Enzymol.* **2,** 113 (1942).
13. M. H. HURST, A. BURGES and P. LATTER, *Phytochem.* **1,** 227 (1962).
14. S. M. SIEGEL, *Physiologia Plantarum* **6,** 134 (1953).

CHAPTER 1

PURIFICATION OF PHENOL OXIDASES

D. S. BENDALL and R. P. F. GREGORY

Department of Biochemistry, University of Cambridge

PHENOL oxidases have been studied since the early work of Bertrand on laccase[1] and mushroom tyrosine oxidase[2] at the end of the last century. The classical work of Kubowitz,[3–5] Keilin and Mann,[6] and Parkinson and Nelson[7] established that these enzymes are copper–protein compounds, contrary to the earlier view of Bertrand[8,9] that manganese was involved. Although much experimental work has been done with them, their mode of action and their biological function in the plant kingdom have remained obscure. However, the last few years have seen a considerable revival of interest, mainly from a purely enzymological point of view, which is the aspect to be considered here. The renewed activity has been stimulated by the development of new techniques both for protein isolation and for studying the mechanisms of the reactions. On the one hand may be mentioned the application of cellulose ion exchangers, which have proved very suitable for chromatography of proteins, and on the other electron spin resonance and exchange of enzyme copper with radioactive copper. At the same time, knowledge of other copper proteins has been developing rapidly. Useful comparisons may be made with substances such as ascorbic oxidase, cytochrome oxidase, plastocyanin, haemocyanin and caeruloplasmin.

An essential stage in the investigation of the properties and mechanism of action of an enzyme is to obtain it in as pure a state as possible; the actual reaction catalysed by the enzyme may remain ambiguous without considerable purification. This paper will discuss the general problems involved in the isolation of phenol oxidases and the properties of these enzymes. The authors' special interest lies with the catechol oxidase occurring in the young green shoots of the tea plant, the enzyme involved in the "fermentation" reaction in the manufacture of black tea. So far as possible, results and experience gained with this enzyme will be used to illustrate the discussion.

PURIFICATION OF ENZYMES

Enzyme purification involves the separation of one protein species from a complex mixture. In a few cases, for example the extracellular laccase produced by some fungi, the enzyme may occur naturally in a relatively pure state, but most phenol oxidases are intracellular enzymes. One may put the problem in perspective by pointing out that about 1000 enzymes are now known. With this figure as a guide, one might expect the amount of one enzyme in a tissue to be, on average, 0·1 per cent of the total protein. In any particular case the purification required to give a pure preparation could be expected to be between 100 and 10,000 times.

While the physical properties of proteins are similar in kind they differ among themselves in degree, and these differences can be exploited.[10–12] Thus, since a high degree of hydration is necessary to render such large molecules soluble in water, proteins tend to be precipitated from solutions of high ionic strength or by high concentrations of organic solvents such as acetone or ethanol. By carefully controlling the addition of salt[13] or organic solvent fractional precipitation of the components of a mixture can often be obtained. In general, the procedure is to divide the material into a series of fractions, each of which is analysed for enzymic activity and protein content. Thus one is able to select a treatment which yields a fraction representing a significant increase in specific activity of the enzyme, with reasonably high recovery. Much of one's time during enzyme purification is spent on these analyses so the methods need to be rapid, at the expense of accuracy if necessary.

The amino acids of any one protein are believed to be specific in number, relative proportions, sequence, and in the degree of concealment or exposure afforded by the tertiary structure. The resulting arrangement of amino acid residues govern fractionation by adsorption and elution employing such media as calcium phosphate gel, alumina and cellulose, while ionizable centres allow ion-exchange chromatography. The application of ion-exchange chromatography to proteins has been immensely successful since the introduction of substituted celluloses by Peterson and Sober.[14,15] The two most widely used for enzyme purification are diethylaminoethylcellulose (DEAE) and carboxymethylcellulose (CMC) in which the charged groups are:

$$\begin{array}{c} \quad\quad\quad\quad\quad\quad\quad C_2H_5 \\ \quad\quad\quad\quad\quad\quad\quad | \\ -O-CH_2-CH_2-N^{+}-H \\ \quad\quad\quad\quad\quad\quad\quad | \\ \quad\quad\quad\quad\quad\quad\quad C_2H_5 \\ \text{DEAE} \end{array} \qquad\qquad \begin{array}{c} \\ \\ -O-CH_2-COO^{-} \\ \\ \\ \text{CMC} \end{array}$$

Their success probably depends on their relatively open structures, allowing the large protein molecule to react with a high proportion of the charged groups of the adsorbent. At the same time the density of the binding sites is not so high that irreversible adsorption occurs nor that the tertiary structure is distorted and the enzyme denatured.

The ionic properties of proteins also allow separations to be made by electrophoresis. Although several systems of electrophoresis of proteins are employed the most useful from a purely preparative point of view is that of Porath[16] in which the material is submitted to zone electrophoresis on a vertical cellulose column. In the starch gel technique of Smithies,[17] which has a particularly high resolution, the mobility depends not only on the electrophoretic effect but also on the ability of the molecule to penetrate the interstices of the solid phase. The latter is known as gel filtration or molecular sieving. Gel filtration by itself can be used to separate proteins employing dextran polymers of varying degrees of cross-linkage ("Sephadex"),[18] and more recently polyacrylamide.[19] Advantage can also be taken of the varying degrees of resistance shown by proteins to heat denaturation. Some selective effect can often be obtained by the presence of the substrate, which stabilizes the enzyme.

To obtain an enzyme in a pure state several of these fractionation procedures need to be used in sequence. The suitability and detailed method of application of any technique has to be determined to a large extent by trial, guided by measurements of the specific activity in the preparations and by knowledge which accumulates during the course of the work concerning the properties of the molecule. Towards the end, criteria are needed to establish the purity of the preparation. Crystallization of a protein is not a satisfactory guide, as protein crystals only 50 per cent pure are well known; it is attempted by biochemists mainly for aesthetic or taxonomic reasons. Physical techniques involving the analytical ultracentrifuge and the Tiselius electrophoretic system are of value in testing the homogeneity of a preparation in addition to their ability to evaluate certain physical constants of the molecule. Solubility measurements are also a sensitive test.[12]

No single criterion can be considered adequate. It is a great help for example to be able to demonstrate, in addition, a constant ratio between light absorption at 280 mμ, due to aromatic amino acid residues, and the enzymic activity as the enzyme is eluted from a chromatographic column. Other chemical criteria such as end-group analysis can also be used.

ASSAY OF PHENOL OXIDASES

Several methods of assaying phenol oxidases have been used, depending on measurement of oxygen uptake in Warburg manometers, formation of coloured products such as purpurogallin[6] or dopachrome[20] or on measurement of the disappearance of ascorbic acid due to coupled oxidation in the presence of a phenolic substrate. In the last case ascorbic acid may be determined spectrophotometrically[21] or by indophenol titration[22] or by detection of the appearance of *o*-quinone with an external indicator.[23] Each method has its own advantages and disadvantages. The authors have chosen to measure the oxygen uptake by means of the Clark oxygen electrode[24,25] using pyrogallol as the substrate at pH 5·6. This technique has the advantage that once the apparatus is set up the experimental manipulations are extremely simple and rapid. It is versatile, as the assay conditions can be easily varied and the system can be coupled to a chart recorder.

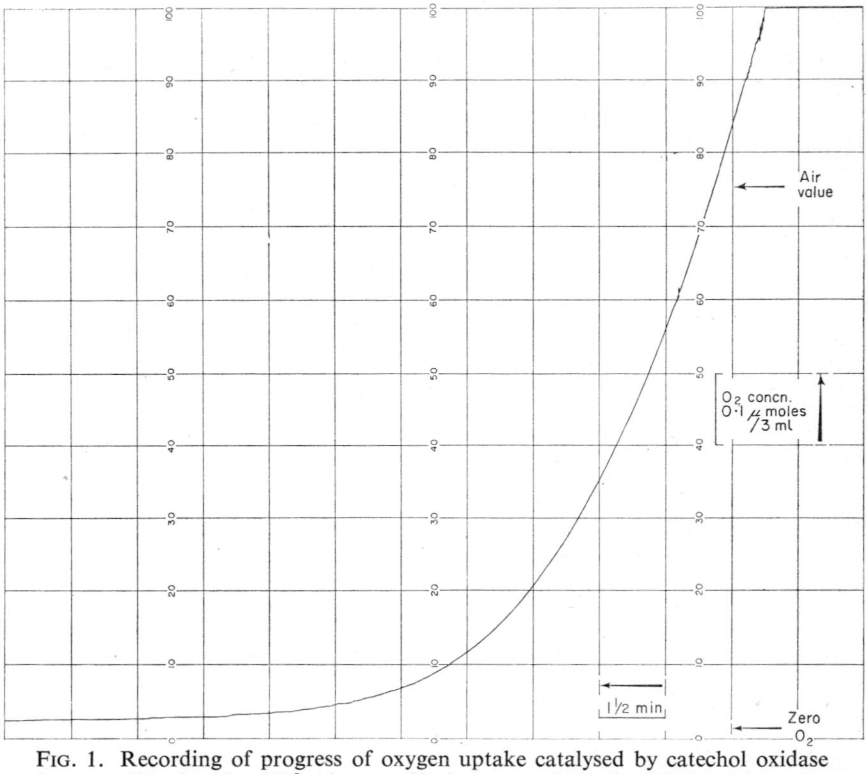

FIG. 1. Recording of progress of oxygen uptake catalysed by catechol oxidase from tea (fraction 5) at 30° using the Clark oxygen electrode. The reaction mixture contained 0·05 M-sodium citrate buffer, pH 5·6, 0·1 M-pyrogallol and enzyme in a total volume of 3 ml.

The electrode depends in principle upon the reduction of oxygen at a polarized platinum cathode of an electrolytic cell in which the other electrode is a silver/silver chloride wire in contact with saturated KCl. A polarizing potential of 600 mV is applied to the electrode, causing a current to flow which can be measured directly with a sensitive galvanometer or transformed into a voltage and fed into a chart recorder. The magnitude of the current is directly proportional to the concentration of oxygen in the medium in which the electrode is suspended.

The Clark oxygen electrode differs from the others in the arrangement of the platinum cathode. In this case it is a small disc lying behind a thin polythene membrane which has a selective permeability towards oxygen. The electrode itself is bathed by a thin film of KCl electrolyte which connects with the reference electrode by an annular ring. Poisoning of the electrode by components of the reaction mixture is prevented by the membrane. The low affinity of phenol oxidases for oxygen precludes a linear progress curve, since the oxygen tension decreases during the reaction. This is shown in Fig. 1 by a recording obtained with the tea enzyme. However, for routine assays a tangent can be drawn at some standard oxygen

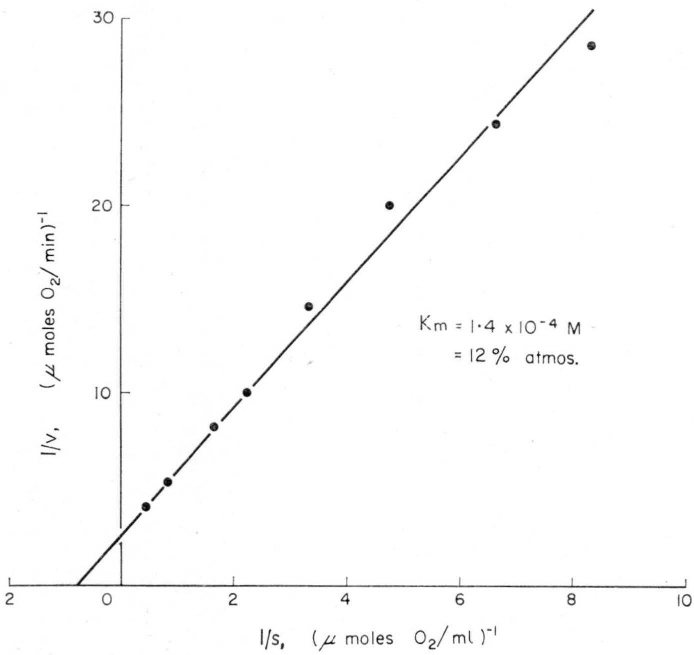

FIG. 2. Lineweaver–Burk plot calculated from the curve in Fig. 1 to show the affinity of the tea leaf catechol oxidase for oxygen.

tension, most conveniently that of air. Also, a single recording, which takes only a few minutes to obtain, provides immediately the data from which the oxygen affinity of the enzyme can be calculated. Thus, by replotting the data on Fig. 1 by the method of Lineweaver and Burk[26] (Fig. 2), the Michaelis constant was found to be $1{\cdot}4 \times 10^{-4}$ M (12 per cent O_2) for the purified tea enzyme.

SPECIAL PROBLEMS IN ISOLATING PHENOL OXIDASES

Although general methods of isolation are similar for any enzyme, purification of a phenol oxidase does involve problems deserving special mention. The first of these is the choice of test substrate. A tissue homogenate will be capable of catalysing the oxidation of several phenolic substrates, and more than one type of phenol oxidase is known. It will be helpful, therefore, to consider what is known of the substrate specificities of the purified enzymes (Table 1). The most general activity is the oxidation of *o*-diphenols to the corresponding *o*-quinones, in fact all known phenol oxidases carry out this type of reaction. Not only may *o*-diphenols be oxidized, but also *p*-diphenols and monophenols. *m*-Diphenols undergo direct enzymic oxidation only rarely, and in some cases they are inhibitory.(27) It has been established, however, that the extracellular inducible laccase produced by the fungus *Polyporus versicolor* will oxidize resorcinol and phloroglucinol.(28) Some substances can be oxidized in different ways according to the type of enzyme; quinol and monophenols are oxidized by *o*-hydroxylation by potato and mushroom enzymes,(29–33) and by a second mechanism, which is probably dehydrogenation to a free radical, by the laccases.(28,34)

Laccases have the broadest specificity range of all the phenol oxidases. Not only do they oxidize *o*- and *p*-diphenols, and sometimes *m*-diphenols and monophenols, but also ascorbic acid(35) and *p*-phenylenediamine. It may be noted in passing that ascorbic oxidase itself has a very narrow specificity range, including besides ascorbic acid only a few closely related ene-diols.(36,37) Recent work on the mode of action of laccase(28,34,35,38,39) suggests that all laccase substrates are oxidized by the same type of mechanism. By contrast, extracts of mushrooms or potatoes clearly oxidize *p*-cresol and catechol in different ways. The relationship between the two reactions and the mechanisms involved have been, and remain, some of the most fundamental problems in this field. However the two activities are widely considered to be due to the same enzyme protein, because they are associated with highly purified, apparently homogeneous preparations.(40–44) Some phenol oxidases have a more limited specificity in that they will only attack *o*-diphenols. *p*-Diphenols and monophenols are completely inactive even in crude preparations. Such enzymes, which might be described as true catechol oxidases, have been found in leaves of tea(45,46) and tobacco(47) and in sweet potatoes.(48)

On the basis of the above, three main classes of phenol oxidase can be distinguished; the catechol oxidase type of tea and tobacco, the cresol or tyrosine oxidase type of mushroom and potato, and the laccases. Comparisons of enzymes within each class reveal finer specificity distinctions. For example, Clayton(47) has described a chlorogenic oxidase in tobacco leaves which appears to be incapable of oxidizing catechol itself. There is some evidence for a similar enzyme in *Solidago* leaves(49) and in both cases it has been suggested that the tissue contains a second enzyme which is able to oxidize catechol. These varied observations show that the choice of

TABLE 1. SPECIFICITIES OF COPPER OXIDASES

	Monophenols		Diphenols					
					p-			
	Hydroxylation	Dehydrogenation	*o*-	*m*-	Hydroxylation	Dehydrogenation	Ascorbic acid	References
Ascorbic oxidase	−	−	−	−	−	−	+	91, 92
Phenol oxidases:								
Tea leaf	−	−	+	−	−	−	−	45, 46
Sweet potato	−	−	+	−	−	−	−	48
Tobacco leaf chlorogenic oxidase	−	−	+	−	−	−	−	47
Potato	+	−	+	−	+	−	−	5, 32, 93
Mushroom	+	−	+	−	+	−	−	6, 30, 41, 42, 44
Lac laccase	−	?	+	−	−	+	+	35, 51, 76
Fungal laccase	−	+	+	+	−	+	+	28
Caeruloplasmin	−	−	+	?	−	+	+	94

test substrate when purifying a phenol oxidase is important and it should be remembered that more than one enzyme may be present in a tissue.

In general, a phenol oxidase is accompanied in a tissue by phenolic substances capable of being oxidized by the enzyme. We assume that enzyme and substrate are spatially separated in the intact tissue, but as soon as the material is disintegrated in order to extract the enzyme the browning reactions begin. The products of oxidation, quinones and condensed products with tanning properties, react with the enzyme protein. The resulting complex may be catalytically inactive (reaction inactivation is common) or there may arise an active protein with modified properties. As an example, the removal of coloured impurities by precipitation with lead acetate, sometimes with the simultaneous loss of a large proportion of the activity, is frequently quoted in the literature. Such tanning effects probably explain the very small yield of colourless enzyme obtained by Keilin and Mann[6] in their classical work with the mushroom enzyme. Evidence that tanned and native enzyme fractions can be separated on DEAE will be discussed below. In most of the recent work the choice has been to prepare an acetone powder as the first step in purification, the enzyme being precipitated with little loss of activity at the same time as most of the endogenous phenols are extracted into the acetone. Alcohol would be a better solvent for the phenols, but usually causes considerable inactivation of the enzyme.

Once a phenol-free residue has been prepared, the next stage is to extract the enzyme in soluble form. The ease with which this can be done varies considerably from one tissue to another. Extraction is most likely to be successful with an aqueous solution of moderately high ionic strength and appropriate pH, but in some cases distilled water[35,50,51] or dilute buffer of neutral pH are effective.[47] Thirty per cent acetone has also been used successfully.[43] This varying ease of extraction may be related to the localization of the enzyme in the intact cell. Phenol oxidases have been found in particle-bound form in homogenates from leaves of spinach-beet,[52] sugar-beet,[53] tobacco[54] and tea,[55] and in lettuce seedlings.[56] As yet the experimental evidence does not adequately identify the particles to which the enzymes are bound, although chloroplasts, grana and mitochondria have all been suggested. On the other hand, the enzyme occurs in soluble as well as particulate form in tobacco and sugar-beet leaves, and the behaviour of the mushroom and potato enzymes suggests that they also are soluble enzymes, although no cell fractionation work seems to have been done. The soluble enzymes should be readily extracted from the tissue, and in some circumstances particulate enzymes may also be fairly easily obtained in soluble form. However, in many cases the enzyme occurs tightly bound to the insoluble structure of the particle, which is largely lipoprotein in nature. Treatment with organic solvents or detergents is likely to render the enzyme more soluble, as Mayer and Friend[57] have found for the enzyme in the chloroplast fraction of sugar-beet leaves. However, resort to digestion by phospholipases or other enzymes may be necessary to obtain the enzyme in truly soluble form.

Another factor of special interest is the masked or latent form in which a phenol oxidase sometimes occurs in freshly prepared homogenates. Several kinds of latency are known with enzymes generally. A common type depends on the presence of the enzyme within some particle which cannot be penetrated by the substrate when supplied externally. The enzyme then appears to be activated by treatments such as freezing and thawing, mechanical disintegration in a Waring Blendor, and exposure to detergents, hypotonic conditions and certain enzymes, all of which damage the membranes of the particle.[58,59] So far, this type of behaviour has not been found with phenol oxidases. Enzymes may also occur as inactive precursors, of which two types may be mentioned; one is activated by proteolytic removal of a peptide fragment of the molecule (pepsinogen, trypsinogen, etc.); the other by treatment with organic solvents, certain detergents, urea, mildly alkaline conditions, or moderately high temperatures, all of which can be expected to modify the tertiary structure of the molecule. Protyrosinases of the latter type have been found in grasshopper eggs,[60,61] mealworms,[62] adult *Drosophila*[63] and crayfish serum.[62] From our present standpoint Kenten's[64,65] discovery of the enzyme of broad-bean leaves in this kind of latent form is specially important. The amount of latent enzyme can easily be determined by carrying out the normal assay in the presence and absence of a suitable detergent such as aerosol OT, sodium dodecylsulphate, or Teepol. Bailey[66] has recently described a small-scale purification of the broad-bean precursor by chromatography on DEAE-cellulose. Activation modified the properties of the protein, as was shown by the fact that the active enzyme was very strongly adsorbed onto DEAE-cellulose.

An apparently low activity may also be due to a naturally-occurring inhibitor. An interesting discovery of this kind has recently been the subject of preliminary reports by Karkhanis and Frieden.[41,67] A protein with inhibitory properties has been found to accompany the mushroom tyrosinase until the last stage of purification when it is separated out by chromatography on DEAE. This inhibitor shows several remarkable properties. After its removal the purified enzyme will oxidize tyrosine without any lag phase. Addition of the inhibitor to the enzyme not only causes a decreased rate of oxidation of tyrosine and 3,4-dihydroxyphenylalanine but also induces a lag in the oxidation of tyrosine. The inhibitor has the rather low copper content of 0·06 per cent, and incubation with cupric ions is claimed to convert the inhibitor into active enzyme.

Similarly, removal of copper from purified tyrosinase was found to produce an apoenzyme which also had inhibitory properties.

PURIFICATION OF CATECHOL OXIDASE FROM TEA

Some of the points which have been raised during this discussion may be illustrated by the authors' work with the catechol oxidase of tea, but it is stressed that this is still in progress and therefore incomplete. A partial

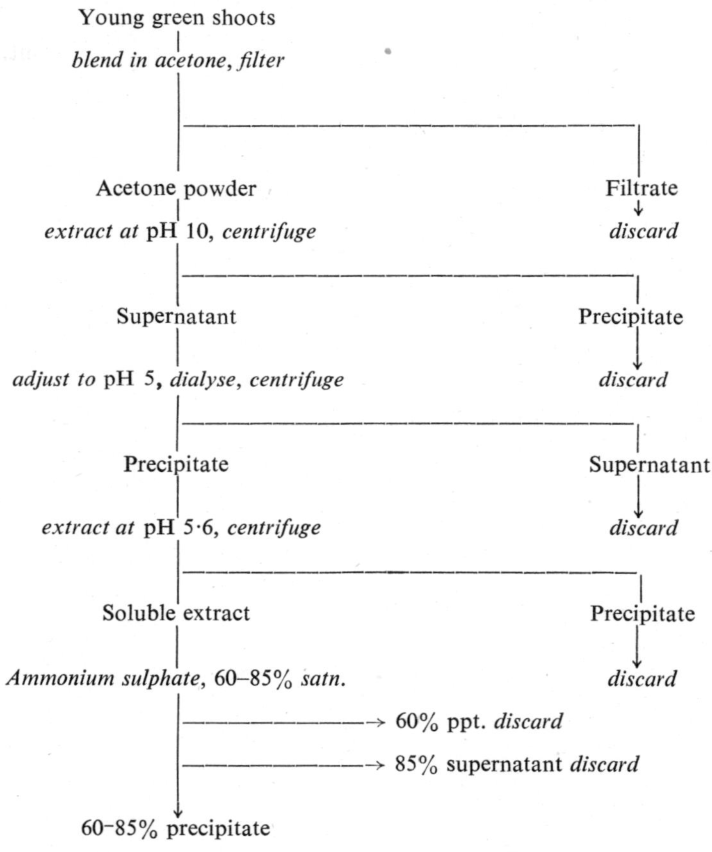

FIG. 3. Flowsheet for the author's procedure for purifying the tea leaf catechol oxidase.

purification of the tea enzyme was achieved by Sreerangachar.(68) In the present work the first two stages are similar to those of Sreerangachar, but thereafter the procedures differ.

The preparation of an acetone powder forms the most satisfactory first stage (Fig. 3). However, with some powders it was found very difficult to extract appreciable amounts of enzyme, although in other cases a 60 per cent extraction was readily obtained with alkaline buffer of high ionic strength. Further work showed that good extraction of the enzyme from the powder depends on the presence of about 20 per cent water in the acetone used for its preparation, except for a final drying with pure acetone. This effect may be explained by the greater solubility of phospholipids in aqueous rather than dry acetone, since such evidence as there is[55] suggests that the tea enzyme is particle-bound; alternatively it may be associated with the lower proportion of residual phenols in the powder prepared with aqueous acetone.

The acetone powder is extracted with glycine buffer, about pH 9·8, containing M-sodium chloride and 0·04 M-ascorbate to reduce oxidation by remaining phenols. Insoluble material is removed by centrifuging, the supernatant adjusted to pH 5 and dialysed against tap water overnight. A flocculent precipitate is formed, consisting mostly of denatured protein, but fortunately all the activity is associated with the precipitate in a form which can readily be extracted with 0·1 M-citrate buffer at pH 5·6. This conveniently concentrates the enzyme at the same time as purifying it about ten times. The solution can then be fractionated with ammonium sulphate. The material precipitating between 60 and 85 per cent saturation contains 90 per cent of the activity.

The precipitate is dissolved in 0·1 M-sodium citrate pH 5·6 and then dialysed against 0·05 M-sodium acetate buffer of the same pH (Fig. 4). A precipitate forms containing much of the activity, but some activity remains in solution. This is a clear-cut separation; if the precipitate is suspended in more buffer and dialysed again, 99 per cent of the activity remains insoluble. When the precipitate is dissolved it shows a relatively high light absorption at 260 mμ, and the presence of ribonucleic acid has been detected by the orcinol reaction. On the other hand the soluble fraction appears to be free from nucleic acid and can be chromatographed on CMC. Two peaks of activity are obtained from this first column. The first fraction (F3) comes through unretarded and is brownish-yellow. A salt gradient elutes fraction F2, which is yellow, followed by peroxidase and an inactive blue protein, but the separations are incomplete. Both fractions F2 and F3 stick firmly to DEAE.

The fraction insoluble on dialysis at pH 5·6 is now dissolved and dialysed further against 0·05 M-tris buffer pH 8·4. This again separates the activity into soluble (F4) and insoluble (F5) fractions. Both fractions are treated with protamine sulphate which precipitates nucleic acid and much coloured material, leaving the activity in solution. These two soluble fractions can now be chromatographed on CMC as before. In the one case the final solution is yellow and the enzyme probably tanned. Some, but not all, of the activity is strongly bound by DEAE. The other fraction, after two runs on CMC, yields a pale blue solution. This blue fraction has the highest specific activity we have yet attained, and on a protein basis represents a purification of about 2000 times over the acetone powder. Under the conditions in which the other fractions stick firmly to DEAE, the blue fraction is not held at all.

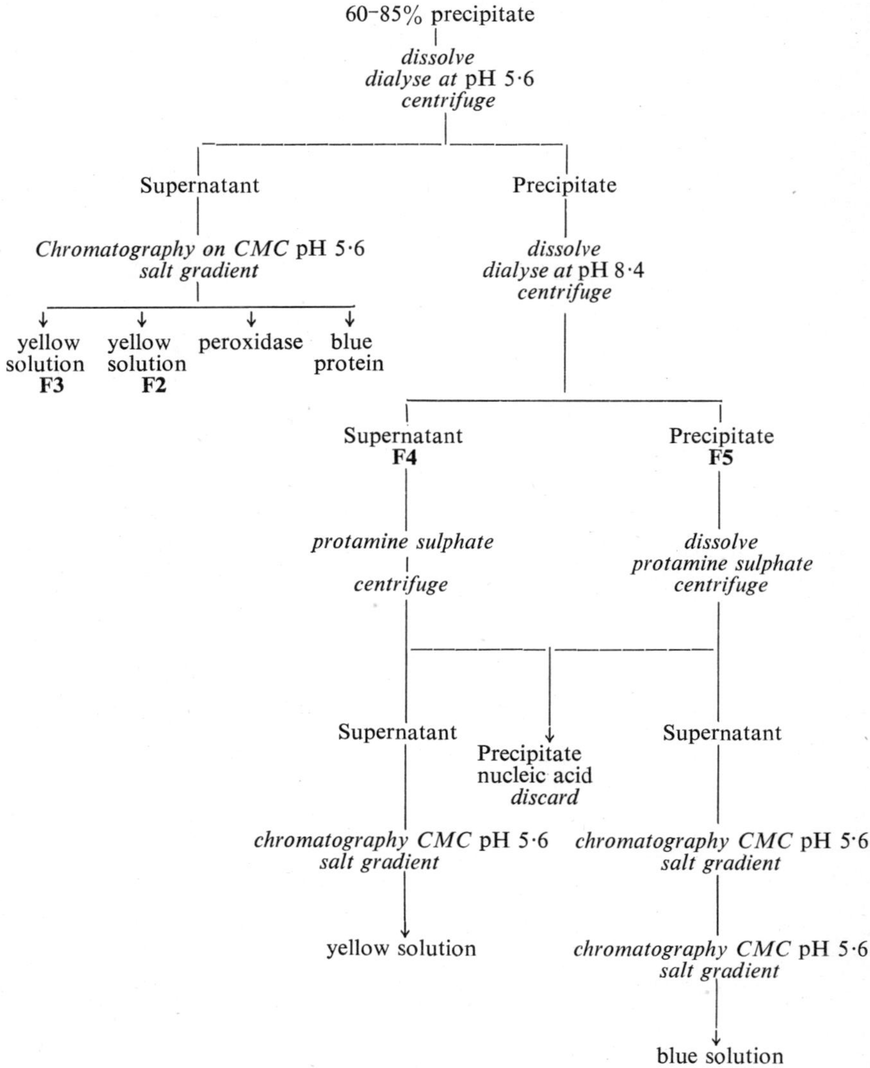

Fig. 4. Flowsheet for the author's procedure for purifying the tea leaf catechol oxidase (*continued*).

Thus, four distinct active fractions have been obtained. Do these represent different enzymes, or are they various modified forms of one protein? The evidence available favours the latter hypothesis, and it is assumed that the blue fraction represents native enzyme, which can be modified in various ways by tanning. In support of this, the blue fraction, when allowed to act on catechol until the solution goes dark brown, is converted into a tanned fraction which is strongly bound by DEAE. The reverse change has not been achieved but it is of some interest that triethylamine acetate prevents F3 from adhering to DEAE. Some, but not all, of the yellow colour is still held by the column while more can be dialysed out of the eluate.

Similar irreversible binding of coloured tyrosine oxidase fractions to DEAE is evident in the work of Frieden and Ottesen[40] and of Brown and Ward.[69]

PROPERTIES AND MODES OF ACTION OF PURIFIED PHENOL OXIDASES

This discussion will be concluded by considering some of the results obtained by studies of the highly purified phenol oxidases from various sources. Some experimental data concerning the copper in phenol oxidases and other copper proteins are collected together in Table 2.

The enzymes can be divided into two classes by colour. The best preparations of the mushroom enzyme are colourless;[6,40] Kubowitz[5] described solutions of the potato enzyme as yellow, but this is suggestive of some degree of tanning of a possibly colourless native enzyme. On the other hand the laccases are blue, including the fungal laccase of Malmström, Fåhraeus and Mosbach,[70] caeruloplasmin, the copper protein of mammalian blood plasma purified by Holmberg and Laurell,[71] and the lac tree latex enzyme. The purified ascorbic oxidase preparation of Dunn and Dawson[72] was green, and it is also interesting to note that blue copper proteins with no oxidase activity have been described in plants. These are the plastocyanin of Katoh and Takamiya[73,74] occurring in *Chlorella* and chloroplasts of higher plants, and a similar substance isolated from lac latex by Omura.[75] The blue colour is associated with the absorption band towards the red end of the spectrum. It is bleached by substrate, lost when the enzyme is denatured, or when the copper is removed by dialysis against cyanide, and restored, together with the activity, by incubating the apoenzyme with copper salts. The blue colour thus seems to be due to the copper–protein complex, probably in the oxidized form.[35,38,71,73,76–81] Data concerning the molecular weights and copper contents of these substances are incomplete, but the oxidases seem to contain at least four copper atoms per molecule. Caeruloplasmin contains eight and ascorbic oxidase six. The presence of four copper atoms may be associated with the fact that oxygen is reduced to water, a four equivalent change, without the accumulation of detectable intermediates. Omura's inactive blue protein from lax latex contains only one copper atom, while plastocyanin contains two. It is of some interest to note that all oxidases

TABLE 2. PROPERTIES OF SOME COPPER PROTEINS

	Colour	Mol. wt.	Cu content (%)	Cu content (atoms/mol.)	Cu valency*	E_0' pH7 (mV)	References
Mushroom phenol oxidase	Colourless	102,000	0·2–0·3	4	+1?		6, 44, 95
Potato phenol oxidase	Yellow?		0·2		+2?		3–5
Chinese lac laccase	Blue	120,000–140,000	0·22–0·25	4·2–5·7	+2	+415	35, 51
Fungal laccase	Blue	57,000	0·4	3·6	+2		19, 70, 77, 83
Caeruloplasmin	Blue	151,000	0·32	7·6	{+2, +1} ?		71,79
Ascorbic oxidase	Green	146,000	0·26	6			72
Plastocyanin	Blue	21,000	0·58	1·9	+2	+370	74
Lac blue protein	Blue	25,000	0·33	1·3			75
Haemocyanin	Purplish-blue or greenish-blue	613,000†	0·18–0·25	25†	+1	+550?	96–99

* In the aerobic state for oxidases and the blue state for other proteins.
† These values vary with the species, and those for squid haemocyanin(97) are quoted as an example.

known to reduce oxygen to water rather than hydrogen peroxide contain copper.

The classical explanation of the dehydrogenation function of phenol oxidases has been that the copper acts as a built-in electron carrier undergoing reversible oxidation and reduction. There has been little direct evidence to support this theory until the recent work of Nakamura[35,82] neatly demonstrated its validity for lac laccase. The light absorption of the enzyme at 615 mμ was proportional to the copper content in various preparations. Measurements of extinction at this wavelength could be used to follow reduction or oxidation of the enzyme copper during titration with substrates or in the presence of redox buffers. With ferrocyanide-ferricyanide mixtures a redox system was revealed involving a one-electron change with a potential of +415 mV at pH 7. Titration with quinol showed that two copper atoms are reduced per molecule of quinol; and titration of leuco-enzyme with oxygen that one molecule of oxygen is involved in the re-oxidation of four atoms of copper. Measurements of the difference in magnetic susceptibility between the oxidized enzyme and either the copper-free apoenzyme or the ascorbate-reduced enzyme revealed the presence of one unpaired electron per copper atom in the oxidized state. This is most readily interpreted by supposing that the four copper atoms of the molecule are cupric in the aerobic state, and reduced to cuprous by substrate. The enzyme–catalysed reaction can therefore be written in the form:

$$2Cu^{++} + \text{Quinol} \rightarrow 2Cu^{+} + p\text{-benzoquinone} + 2H^{+}$$

$$4Cu^{+} + O_2 + 4H^{+} \rightarrow 4Cu^{++} + 2H_2O.$$

Malmström's group have also been able to show the presence of cupric copper in the fungal laccase and caeruloplasmin[77,83] by electron spin resonance spectrometry. With the fungal enzyme a reduction of the signal by substrate in the aerobic steady state was demonstrated, which indicated reduction of copper from cupric to cuprous by substrate.

Our knowledge of the state and function of the copper in other phenol oxidases is too scanty for any but speculative comparisons to be made at present. However, important work by Nakamura[34] and Mason, Yamazaki and co-workers[84–86] has been recently published in which the mechanism of substrate oxidation has been studied by methods involving electron spin resonance. Two types of mechanism have been revealed. With laccase, ascorbic oxidase and peroxidase the primary product of the enzymic reaction is the free radical produced by removal of one hydrogen atom from the substrate; the radical then undergoes dismutation rather than further enzymic oxidation. With the mushroom enzyme, by contrast, no free radical can be detected as the immediate result of the enzymic oxidation of catechol, and the first product is *o*-quinone produced by a two-equivalent transfer.

These results bring together several scattered observations. They support Freudenberg's[87] theory for the mechanism of oxidative formation of lignin-like polymers from coniferyl alcohol in model systems, and they also explain why, as found by Higuchi,[88,89] laccase and peroxidase are active in this system while mushroom tyrosinase is not. They also explain why fungal laccase and peroxidase give similar products of oxidation of *p*-cresol,[28] which differ from the *o*-diphenol obtained with the mushroom and potato enzymes.

Enough has been said to show that while rapid progress is now being made in the study of phenol oxidases with the help of new experimental tools, much still remains to be learned. It seems appropriate to end with a plea for the comparative approach. The distribution of catechol oxidases in higher plants was studied by Mrs. Onslow in the early 1920's.[90] She found them in 63 per cent of the Angiosperm families; it is sufficient to comment that a mere handful of species have been studied in detail, and yet remarkably varied properties have been discovered.

Acknowledgements

This work has been supported by the Nyasaland Tea Association, the Tea Research Institute of East Africa, and the Agricultural Research Council. We wish to thank Dr. R. Hill, F.R.S., and Professor F. G. Young, F.R.S., for their interest in the work, and Dr. C. V. Cutting for supplying tea leaf material.

REFERENCES

1. G. BERTRAND, *C.R. Acad. Sci., Paris* **118,** 1215 (1894).
2. E. BOURQUELOT and G. BERTRAND, *C.R. Soc. Biol., Paris* **47,** 582 (1895).
3. F. KUBOWITZ, *Biochem. Z.* **292,** 221 (1937).
4. F. KUBOWITZ, *Biochem. Z.* **296,** 443 (1938).
5. F. KUBOWITZ, *Biochem. Z.* **299,** 32 (1938).
6. D. KEILIN and T. MANN, *Proc. Roy. Soc. B* **125,** 187 (1938).
7. G. G. PARKINSON and J. M. NELSON, *J. Amer. Chem. Soc.* **62,** 1693 (1940).
8. G. BERTRAND, *Bull. Soc. Chim. Paris* (3) **17,** 577 (1894).
9. G. BERTRAND, *C.R. Acad. Sci., Paris* **124,** 1032, 1355 (1897).
10. M. DIXON and E. C. WEBB, *Enzymes*, Longmans, Green, London (1958).
11. S. SCHWIMMER and A. B. PARDEE, *Advanc. Enzymol.* **14,** 375 (1953).
12. J. F. TAYLOR in *The Proteins*, Vol. 1a, p. 1, eds. H. Neurath and K. Bailey, Academic Press, New York (1953).
13. M. DIXON and E. C. WEBB, *Advanc. Protein Chem.* **16,** 197 (1961).
14. E. A. PETERSON and H. A. SOBER, *J. Amer. Chem. Soc.* **78,** 751 (1956).
15. E. A. PETERSON and H. A. SOBER in *Methods in Enzymology*, Vol. 5, p. 3, eds. S. P. Colowick and N. O. Kaplan, Academic Press, New York (1962).
16. J. PORATH, *Biochim. Biophys. Acta* **22,** 151 (1956).
17. O. SMITHIES, *Advanc. Protein Chem.* **14,** 65 (1959).

18. J. Porath and P. Flodin, *Nature, Lond.* **183,** 1657 (1959).
19. S. Hjertén and R. Mosbach, *Analyt. Biochem.* **3,** 109 (1962).
20. A. S. Fox and J. B. Burnett, *Proc. Soc. Exp. Biol., N.Y.* **98,** 110 (1958).
21. M. Ashraf and E. Frieden, *J. Amer. Chem. Soc.* **79,** 4854 (1957).
22. H. B. Sreerangachar, *Biochem. J.* **37,** 653 (1943).
23. C. R. Dawson and R. J. Magee in *Methods in Enzymology*, Vol. 2, p. 817, eds. S. P. Colowick and N. O. Kaplan, Academic Press, New York (1955).
24. J. B. Chappell in *Biological Structure and Function, Proc. First IUB/IUBS International Symposium*, Stockholm, Vol. 2, p. 71, eds. T. W. Goodwin and O. Lindberg, Academic Press, New York (1961).
25. P. W. Davies in *Physical Techniques in Biological Research*, Vol. 4, p. 137, ed. W. L. Nastuk, Academic Press, New York (1962).
26. H. Lineweaver and D. Burk, *J. Amer. Chem. Soc.* **56,** 658 (1934).
27. G. Schneider and S. Schmidt, *Hoppe-Seyl Z.* **315,** 20 (1959).
28. G. Fåhraeus and H. Ljunggren, *Biochim. Biophys. Acta* **46,** 22 (1961).
29. W. C. Evans and H. S. Raper, *Biochem. J.* **31,** 2155 (1937).
30. D. C. Gregg and J. M. Nelson, *J. Amer. Chem. Soc.* **62,** 2510 (1940).
31. L. P. Kendal, *Biochem. J.* **44,** 442 (1949).
32. G. Legrand, J. Neumann and G. Lehongre, *C.R. Acad. Sci., Paris* **252,** 2023 (1961).
33. H. S. Mason, W. L. Fowlks and E. Peterson, *J. Amer. Chem. Soc.* **77,** 2914 (1955).
34. T. Nakamura, *Biochem. Biophys. Res. Commun.* **2,** 111 (1960).
35. T. Nakamura, *Biochim. Biophys. Acta* **30,** 44 (1958).
36. M. L. Dodds, *Arch. Biochem.* **18,** 51 (1948).
37. S. W. Johnson and S. S. Zilva, *Biochem. J.* **31,** 1366 (1937).
38. T. Nakamura, *Biochim. Biophys. Acta* **30,** 538 (1958).
39. T. Nakamura, *Biochim. Biophys. Acta* **30,** 640 (1958).
40. E. Frieden and M. Ottesen, *Biochim. Biophys. Acta* **34,** 248 (1959).
41. Y. Karkhanis and E. Frieden, *J. Biol. Chem.* **236,** PC 1 (1961).
42. D. Kertész and O. Azzopardi, *Bull. Soc. Chim. Biol., Paris* **42,** 945 (1960).
43. D. Kertész and R. Zito, *Nature, Lond.* **179,** 1017 (1957).
44. M. F. Mallette and C. R. Dawson, *Arch. Biochem.* **23,** 29 (1949).
45. J. Lamb and H. B. Sreerangachar, *Biochem. J.* **34,** 1472 (1940).
46. E. A. H. Roberts and D. J. Wood, *Biochem. J.* **47,** 175 (1950).
47. R. A. Clayton, *Arch. Biochem. Biophys.* **81,** 404 (1959).
48. I. Z. Eiger and C. R. Dawson, *Arch. Biochem.* **21,** 194 (1949).
49. O. Björkman and P. Holmgren, *Physiol. Plant.* **13,** 582 (1960).
50. M. F. Mallette, S. Lewis, S. R. Ames, J. M. Nelson and C. R. Dawson, *Arch. Biochem.* **16,** 283 (1948).
51. T. Omura, *J. Biochem. Tokyo* **50,** 264 (1961).
52. D. I. Arnon, *Plant Physiol.* **24,** 1 (1949).
53. A. M. Mayer and J. Friend, *J. Exp. Bot.* **11,** 141 (1960).
54. J. H. McClendon, *Amer. J. Bot.* **40,** 260 (1953).
55. L. P. Li and J. Bonner, *Biochem. J.* **41,** 105 (1947).
56. A. M. Mayer, *Physiol. Plant.* **14,** 322 (1961).
57. A. M. Mayer and J. Friend, *Nature, Lond.* **185,** 464 (1960).
58. H. Beaufay and C. de Duve, *Biochem. J.* **73,** 604 (1959).
59. D. S. Bendall and C. de Duve, *Biochem. J.* **74,** 444 (1960).
60. J. H. Bodine and T. H. Allen, *J. Cell. Comp. Physiol.* **11,** 409 (1938).
61. J. H. Bodine and T. H. Allen, *J. Cell. Comp. Physiol.* **12,** 71 (1938).
62. J. H. Bodine and T. H. Allen, *J. Cell. Comp. Physiol.* **18,** 151 (1941).
63. N. H. Horowitz and M. Fling in *Amino Acid Metabolism*, p. 207, eds. W. D. McElroy and B. Glass, Johns Hopkins Press, Baltimore (1955).
64. R. H. Kenten, *Biochem. J.* **67,** 300 (1957).
65. R. H. Kenten, *Biochem. J.* **68,** 244 (1958).
66. J. L. Bailey, *Biochem. J.* **79,** 514 (1961).
67. Y. Karkhanis and E. Frieden, *Biochem. Biophys. Res. Commun.* **4,** 303 (1961).
68. H. B. Sreerangachar, *Biochem. J.* **37,** 667 (1943).

69. F. C. BROWN and D. N. WARD, *J. Biol. Chem.* **233**, 77 (1958).
70. G. MALMSTRÖM, G. FÅHRAEUS and R. MOSBACH *Biochim. Biophys. Acta* **28**, 652 (1958).
71. C. G. HOLMBERG and C.-B. LAURELL, *Acta Chem. Scand.* **2**, 550 (1948).
72. F. J. DUNN and C. R. DAWSON, *J. Biol. Chem.* **189**, 485 (1951).
73. S. KATOH, *Nature, Lond.* **186**, 533 (1960).
74. S. KATOH, I. SHIRATORI and A. TAKAMIYA, *J. Biochem. Tokyo* **51**, 32 (1962).
75. T. OMURA, *J. Biochem. Tokyo* **50**, 394 (1961).
76. D. KEILIN and T. MANN, *Nature, Lond.* **143**, 23 (1939).
77. B. G. MALMSTRÖM, R. MOSBACH and T. VÄNNGÅRD, *Nature, Lond.* **183**, 321 (1959).
78. T. OMURA, *J. Biochem. Tokyo* **50**, 389 (1961).
79. S. OSAKI, *J. Biochem. Tokyo* **48**, 190 (1960).
80. A. TISSIÈRES, *Nature, Lond.* **162**, 340 (1948).
81. A. TISSIÈRES, *Nature, Lond.* **163**, 480 (1949).
82. T. NAKAMURA, *Biochim. Biophys. Acta* **42**, 499 (1960).
83. B. G. MALMSTRÖM and T. VÄNNGÅRD, *J. Mol. Biol.* **2**, 118 (1960).
84. H. S. MASON, E. SPENCER and I. YAMAZAKI, *Biochem. Biophys. Res. Commun.* **4**, 236 (1961).
85. I. YAMAZAKI, H. S. MASON and L. H. PIETTE, *J. Biol. Chem.* **235**, 2444 (1960).
86. I. YAMAZAKI and L. H. PIETTE, *Biochim. Biophys. Acta* **50**, 62 (1961).
87. K. FREUDENBERG, *Nature, Lond.* **183**, 1152 (1959).
88. T. HIGUCHI, *Physiol. Plant.* **10**, 356 (1957).
89. T. HIGUCHI, *J. Biochem. Tokyo* **45**, 515 (1958).
90. M. W. ONSLOW, *Biochem. J.* **15**, 107 (1921).
91. S. W. JOHNSON and S. S. ZILVA, *Biochem. J.* **31**, 438 (1937).
92. H. TAUBER, I. S. KLEINER and D. MISHKIND, *J. Biol. Chem.* **110**, 211 (1935).
93. J. NEUMANN, G. LEGRAND, G. LEHONGRE and J. LAVOLLAY, *C.R. Acad. Sci., Paris* **251**, 3091 (1960).
94. C. G. HOLMBERG and C.-B. LAURELL, *Acta Chem. Scand.* **5**, 476 (1951).
95. D. KERTÉSZ, *Nature, Lond.* **180**, 506 (1957).
96. J. B. CONANT, B. F. CHOW and E. B. SCHOENBACH, *J. Biol. Chem.* **101**, 463 (1933).
97. T. OMURA, T. FUJITA, F. YAMADA and S. YAMAMOTO, *J. Biochem. Tokyo* **50**, 400 (1961).
98. A. C. REDFIELD in *Copper Metabolism*, p. 174, eds. W. D. McElroy and B. Glass, Johns Hopkins Press, Baltimore (1950).
99. T. NAKAMURA and H. S. MASON, *Biochem. Biophys. Res. Commun.* **3**, 297 (1960).

CHAPTER 2

PLANT POLYPHENOL OXIDASES AND THEIR RELATION TO OTHER OXIDASE SYSTEMS IN PLANTS

J. G. BOSWELL

Department of Botany, The University, Sheffield

EARLY workers have established that about 60 per cent of the Phanerogams contain phenolic compounds with the *o*-dihydroxy group and that these tissues showed a characteristic browning when cut and the surface exposed to air. Following the work of Bach and Chodat, Onslow[1] reported the results of an extensive survey of the Angiosperms which determined the distribution of the enzyme which Bach and Chodat called oxygenase and which is now known in more general terms as a phenolase. She noted that the two major divisions of the Dicotyledons, the Archichlamydeae and the Sympetalae showed different distributions of oxygenase. In the florally more highly developed phylum, the Sympetalae, there was a much wider distribution of the enzymic system, in fact with the exception of the Primulaceae and the Campanulaceae, the Sympetalae included almost wholly oxygenase-containing orders. The Archichlamydeae which are regarded as florally less highly developed, contained a number of orders such as the Crassulaceae, Tropaeolaceae, Betulaceae, Fagaceae and Tiliaceae whose genera did not include oxygenase containing species. While, in general, oxygenase is found in most, if not all, genera of an order, there are orders such as the Leguminosae in which oxygenase and non-oxygenase containing genera occur, and this is true also of the Ranunculaceae, Liliaceae and the Saxifragaceae. However, it may be stated in very general terms that the more florally advanced group of the Sympetalae includes a greater number of oxygenase containing orders than the Archichlamydeae. The validity of some of these observations may require checking in the light of the known inhibition of phenolases by phenolic compounds.

This was the position when in 1938, Geoffrey Whiting and I,[2] added catechol to thin slices of potato tuber tissue suspended in phosphate buffer and found a substantial stimulation both in oxygen uptake and carbon dioxide output. After the initial stimulation, in which the oxygen uptake reached its maximum before that of the carbon dioxide output, the rates decreased and ultimately there was a substantial degree of inhibition of the respiratory activity of the tissue. Later, we added a crude preparation of polyphenols from the potato tuber itself, purification after alcohol extraction being by the formation of a lead salt. Respiratory stimulation followed the addition of the mixture and no inhibition developed at any stage. This work was then continued, Boswell,[3] by the use of chemically pure substances such as caffeic acid, *p*-cresol, DOPA, protocatechuic acid and other *o*-dihydroxy phenols. It was clear that only catechol and *p*-cresol had inhibitory effects following the initial stimulation. In other cases and particularly with caffeic acid the respiration rate was stimulated and the stimulation continued for long periods, in excess of 7 hr after addition. In particular, the carbon dioxide output was stimulated as well as the oxygen uptake. One peculiarity emerged, that was that fresh cut slices did not oxidize protocatechuic acid whereas slices which had been washed for one or two days in running water with aeration did oxidize this compound (Table 1). To examine this point further the ordered respiratory mechanism of tissue slices was destroyed by freezing at $-20°C$ and thawing in a phosphate buffer before treatment with polyphenols. The phenolase liberated from fresh tissue by such treatment while able to oxidize caffeic acid did not oxidize protocatechuic acid. If, however, the tissue were washed for even 24 hr before use then both substrates were oxidized. It was further shown that if glutamic acid were added to the tissue before the addition of the caffeic acid, then the level of respiratory stimulation was much greater than in the presence of caffeic acid alone. This was in keeping with the observation that crude enzyme preparations in the presence of a *o*-dihydroxy compound would oxidatively deaminate amino acids. I, therefore, put forward the view that the polyphenol oxidase system acted as a terminal oxidase in the respiratory activity of washed potato slices. Then followed the work of Baker and Nelson[4] using protocatechuic acid, Link *et al.*[5] with slices of tomato stem, Hackney[6] with apple tissue and Lerner[7] with the leaves of ivy. Together with Arnon[8] they supported the view that a phenolase acted as a terminal oxidase.

This theory was attacked by those workers who approached the problem from another angle. Initially, due to the work of Schade *et al.*[9] who used respiratory inhibitors, in particular carbon monoxide, and later contributions from a number of workers including Li and Bonner,[10] Stenlid[11] Thiman *et al.*[12] and Hackett *et al.*[13] there developed a school of thought which held that the phenolase system had no part in respiratory mechanism as a terminal oxidase, and that the terminal oxidase in potato tissue and similar tissues was based solely on cytochrome. These workers showed that carbon monoxide inhibited some portion of tissue respiration usually about 50 per cent and that in most cases, except the apple for example, this inhibition was light reversible indicating that an iron containing system of the cytochrome type was involved.

We, therefore, have reached the point at which there are two views of the significance of the phenolase system in tissue respiration. These two views are based on two quite separate groups of experiments, those which involve respiratory stimulation using polyphenols and those which

TABLE 1. EFFECT OF ADDITION OF PHENOLS ON THE RATES OF OXYGEN UPTAKE AND CARBON DIOXIDE OUTPUT BY POTATO TUBER TISSUE

Tissue and treatment	Duration (min)	Excess gaseous exchange (μl.)	
		O_2	CO_2
Fresh cut slices in a phosphate buffer			
Protocatechuic acid*	90	No excess	
Caffeic acid*	90	78	16
Fresh cut slices, frozen at −20°C, suspended in a phosphate buffer			
Protocatechuic acid*	100	No excess	
Caffeic acid*	100	86	0
Slices washed 24 hr, suspended in a phosphate buffer			
Protocatechuic acid*	160	79	76
Slices washed 24 hr, frozen at −20°C, then suspended in a phosphate buffer			
Protocatechuic acid*	140	46	29
Slices washed 48 hr, suspended in a phosphate buffer			
Protocatechuic acid*	200	256	278
Caffeic acid*	200	191	135
Slices washed 48 hr, frozen at −20°C, then suspended in a phosphate buffer			
Protocatechuic acid*	200	120	74
Caffeic acid*	200	51	4

* 1 mg of each phenol was added to about 1 g fresh weight of tissue suspended in 4 ml buffer, 0·02 M phosphate.

involve respiratory inhibition using known respiratory inhibitors. It is useful to consider the exact conditions under which the two groups of experiments are carried out. In both cases, work has dealt with fresh and washed slices of tissue. The results of the washing are respiratory stimulation and if we are dealing with potato tuber tissue, a marked increase during the washing procedure of the protein content of the tissue. We are dealing here with whole cells, the debris of the cells broken by cutting having been removed during the washing period. The tissue shows a high rate of respiration and a marked capacity for protein synthesis, or if you prefer it, shows a net gain in protein content during the washing process. Under the conditions prevailing, the polyphenols in the tissue are in a reduced state, although with progressive washing some degree of permanent oxidation and polymerization occurs since, particularly around the phloem elements, the brown coloured products of oxidation tend to accumulate giving the slices an overall brownish appearance. This maintenance of the polyphenols in the reduced form, which can be demonstrated by isolation and chromatography, means one of two things. If the polyphenol and the phenolase are in contact then there must be present a continuous supply of a reducing agent, e.g. ascorbic acid, which reduces the *o*-quinone resulting from the oxidation of the polyphenol, or the phenolase and the polyphenol are not in contact in the whole cell during this normal metabolic activity. On addition of polyphenol to the solution surrounding the slices, the polyphenol diffuses into the free space of the tissue, and this means in part into the cytoplasm, and there comes in contact with the phenolase system and undergoes rapid oxidation. This oxidation is accompanied by carbon dioxide output and the process appears to involve a cyclic redox system in the case of caffeic acid, the oxygen uptake at the end of 7 hr being far in excess of the theoretical value of half a molecule of oxygen for each molecule of caffeic acid added. Some form of cyclic oxidation-reduction system must occur with some polyphenols when they are added to the external medium. A quite different state of affairs may exist in normal tissue slices. It may be that the naturally occurring polyphenols and the phenolase are spatially separated although this is difficult to envisage since added polyphenols reach the enzyme system so rapidly. It is conceivable, of course, that the naturally occurring polyphenols are bound and inactivated on the non-mitochondrial parts of the cytochrome, or located in the vacuole.

The difficulties with regard to the interpretation of the evidence secured by the use of inhibitors such as carbon monoxide are twofold. Firstly, there is the point that an inhibitor will only produce an inhibition of a multi-stage reaction such as respiration if the inhibitor controls a stage which either is, or becomes under the influence of the inhibitor, a rate limiting reaction. It would appear from the results cited above that part at least of the respiratory mechanism involves the cytochrome system since the carbon monoxide inhibition is light reversible. There is, however, the question of the identity of the carbon monoxide resistant part of the respiratory activity. Since the amount of the phenolase in potato tissue slices is large it may well be that under the conditions of the experiments using carbon monoxide, the activity of the phenolase is not reduced to the level at which it limits the rate of the respiration process. On this view the carbon monoxide insensitive portion of the total respiratory activity would involve the phenolase system as a terminal oxidase. Secondly, the phenolase and cytochrome systems may be alternative terminal oxidases for at least a substantial part of the respiratory mechanism, since both have been linked *in vitro* with the same hydrogen donors. If one is inhibited to a greater or lesser degree then the other may take over either wholly or in part, the role played by the other.

Now it appeared to us that this problem could not be solved by pursuing the two patterns of the earlier experiments. We have, therefore, embarked upon another series of experiments, firstly on the distribution of phenolase and phenolic compounds in the cells and secondly on the biochemical changes which occur in slices of potato tuber tissue through treatment with respiratory stimulants and inhibitors.

Experiments show that potato mitochondria after repeated washing possess considerable phenolase activity, oxidizing protocatechuic and caffeic acids and catechol. The forms of the oxygen uptake-time curves with these three substances are all different. Further the oxidation of catechol does not lead to any inhibition of the enzyme and that of caffeic acid does not give any corresponding output of carbon dioxide. The oxidation of these polyphenols is not stimulated by the addition of cytochrome C. The location of an active phenolase on the mitochondria raises the question as to whether there are two phenolase locations, one on the mitochondria and the other in the cytoplasm or whether the soluble oxidase in the supernatant from high speed centrifugation has been liberated by the breakdown of mitochondria. Chromatographic examination of an ethanol extract of washed mitochondria shows the presence of at least three aromatic substances of which one or two have an *o*-dihydroxy structure.

We have examined the general metabolic activities of potato slices with particular reference to the distribution of nitrogen and phosphate among the fractions.

The phenols used in these experiments may be divided into two groups: firstly, catechol which after initial stimulation of respiration produces a substantial inhibition and secondly, caffeic and protocatechuic acids and DOPA which produce a stimulation for the duration of the experiment, i.e. 3 hr. The phosphate analyses recorded in Tables 2 and 3 are based upon a simple scheme. The dry tissue was extracted with 10 per cent trichloracetic acid, the solution contained the fraction referred to as "acid sol. P", the inorganic phosphate (inorg. P) was determined on this extract. The acid extracted residue gave a value for "acid insol. P". The total phosphate, "total P", was determined on a separate fraction of the dry tissue. The fraction "acid sol. org. P" is the difference between the acid soluble phosphate and the inorganic phosphate. It is not yet possible to interpret the results in Table 2 in detail but it is clear that the addition of these phenols does produce substantial changes in the distribution of phosphate. Catechol modifies the acid sol. P and insol. P ratio and the acid sol. organic P/acid insol. P ratio to some extent by increasing the organic acid soluble fraction but to a much larger extent by bringing about the almost total elimination of the acid insol. P fraction; further

TABLE 2. EFFECT ON PHOSPHATE DISTRIBUTION OF ADDED PHENOLICS (1 MG/ML IN 0·02 M-PHOSPHATE BUFFER; 3 HR) ON POTATO TUBER SLICES WASHED FOR 2 DAYS

Subsequent treatment	Tap water for 3 hr	No further addition	Caffeic acid	Proto-catechuic acid	DOPA	Catechol
Analyses (mg *P*/g D.W.)						
Total P	2·56	4·32	4·01	4·66	4·69	5·76
Acid sol. P	2·00	3·88	3·50	4·05	4·08	5·67
Acid insol. P	0·56	0·44	0·51	0·61	0·61	0·09
Inorg. P	1·50	2·86	2·78	2·94	3·49	4·08
Acid sol. organic P	0·50	1·02	0·72	1·11	0·59	1·59
Acid sol. organic P / Acid insol. P	0·89	2·37	1·41	1·82	0·97	17·2
Acid sol. P / Acid insol. P	3·57	8·81	6·88	6·65	6·70	63·0

TABLE 3. EFFECTS OF KNOWN RESPIRATORY INHIBITORS (1 MG/ML) IN 0·02 M-PHOSPHATE BUFFER; 3 HR) ON POTATO TUBER SLICES WASHED FOR 2 DAYS

Subsequent treatment	No further addition	"Dieca"	8-Hydroxy quinoline	2,4-Dinitrophenol
Analyses (mg P/g D.W.)				
Total P	3·56	3·95	4·39	3·30
Acid sol. P	3·02	3·56	3·81	3·03
Acid insol. P	0·54	0·39	0·58	0·27
Inorg. P	2·12	2·10	2·58	2·20
Acid sol. P (organic)	0·90	1·46	1·23	0·83
Acid sol. P / Acid insol. P	5·58	9·12	6·55	11·2
Acid sol. Org. P / Acid insol. P	1·66	3·74	2·13	3·06

the catechol increases the total P and the inorg. P contents markedly. The effects of catechol are quite different from those of the other three phenols, the non-inhibitory ones. With these, while the inidividual changes in the fractions are variable, the ratios between the fractions follow a pattern, and all result in a reduction of the acid sol. P/acid insol. P and acid. sol. organic P/acid insol. P ratios. How this is brought about has yet to be worked out. These effects have been achieved with all the varieties of potatoes used and at all seasons. It may be noted from Table 3 that

TABLE 4. EFFECT ON NITROGEN DISTRIBUTION OF ADDED PHENOLICS (1 MG/ML IN 0·02 M-PHOSPHATE BUFFER; 3 HR) ON POTATO TUBER SLICES WASHED FOR 2 DAYS

Subsequent treatment	No further addition	Catechol	Protocatechuic acid	DOPA	Caffeic acid
Analyses					
Total N (mg /g D.W.)	15·17	12·15	14·51	14·69	15·30
Alcohol insol. N (protein)	8·31	7·73	8·87	9·24	6·70
Amino acid N	2·62	1·31	2·93	2·93	3·59
Stable amide N	0·78	0·53	0·55	0·61	0·75
Labile amide N	1·18	0·68	1·13	1·05	1·45
Ammonia-N	0·27	0·30	0·34	0·33	0·28
Stable/Labile amide	0·66	0·78	0·49	0·58	0·52
Total amide/Amino acid N	0·75	0·92	0·57	0·57	0·61
Protein N/Total N	0·55	0·64	0·61	0·63	0·44

known respiratory inhibitors such as "Dieca" (sodium diethyl-dithiocarbamate), 8-hydroxyquinoline and 2,4-dinitrophenol produce changes in the ratios of phosphate fractions similar to those produced by the inhibitor catechol and unlike those of the other phenolics. We are attempting to carry this investigation further using paper chromatography but have found very considerable difficulties in securing quantitative analyses. The quantitative separation of a wide range of organic phosphates on paper in single solvent systems has proved impossible. We hope to solve this technical difficulty by using high voltage paper electrophoresis.

Turning now to the changes which have occurred in the distribution of nitrogen in a parallel series of analyses. The total nitrogen was determined on a sample of dried tissue, total N. A separate sample was extracted with 80 per cent ethanol and the following values were determined on this fraction: total soluble N, amino acid N, stable and labile amide N and ammonia-N. The analysis in Table 4 is on tissue which gave the phosphate pattern in Table 2, that is with the very large increase in phosphate following treatment with catechol. The marked feature is the fall in total nitrogen following treatment with catechol, this is largely at the expense of the soluble nitrogen and is in marked contrast to the accumulation of phosphate under the same experimental conditions. The bulk of the loss of nitrogen is in the form of ammonia from deamination within the tissue. Caffeic acid alone produces a fall in protein N as a fraction of the total N and this is balanced by an increase in the amino acid and amide fractions. When, however, the ratios between various soluble fractions are examined then catechol with its inhibitory action stands apart from the other phenolics which stimulate respiration.

We have here evidence of metabolic changes occurring when *o*-dihydroxy phenols are applied to the tissue. Again seasonal and varietal patterns have been found and the full significance of these results can only be appreciated after more detailed analysis of the fractions, particularly of the amino acids.

In conclusion, the position may be summarized in this way, the initial groups of experiments led to a stalemate since both provided data from which unassailable conclusions could not be drawn. The solution has to be sought along other lines: firstly, a study of the enzyme complex using pure preparations to determine the properties of the system and if possible to provide highly specific inhibitors, this is the work of enzymologists; secondly, to determine the location of the enzyme and substrates in the cell and in this we have made some progress; finally, to observe changes in intermediates following control of the overall respiratory rates both by stimulation and inhibition. I have provided a brief report of the mass of elementary data which we have accumulated in this field, but while the work is promising it is far from providing a solution to the problem.

REFERENCES

1. M. W. ONSLOW, *The Principles of Plant Biochemistry*, Cambridge University Press (1931).
2. J. G. BOSWELL and G. C. WHITING, *Ann. Bot.* **2,** 847, new series (1938).
3. J. G. BOSWELL, *Ann. Bot.* **9,** 55, new series (1945).
4. D. BAKER and J. M. NELSON, *J. Gen. Physiol.* **26,** 269 (1943).
5. G. K. K. LINK, R. M. KLEIN and E. S. G. BARRON, *J. Exp. Bot.* **3,** 216 (1952).
6. F. M. V. HACKNEY, *Proc. Linnean Soc. N.S. Wales* **73,** 439 (1948).
7. N. H. LERNER, *J. Exp. Bot.* **5,** 79 (1954).
8. D. I. ARNON, *Copper Metabolism*, a symposium, Johns Hopkins, Baltimore (1950).
9. A. L. SCHADE, L. BERGMANN and A. BYER, *Arch. Biochem.* **18,** 85 (1948).
10. P. L. LI and J. BONNER, *Biochem. J.* **41,** 105 (1947).
11. G. STENLID, *Physiol. Plant* **2,** 61 (1949).
12. K. V. THIMANN, C. S. YOCUM and D. P. HACKETT, *Arch. Biochem.* **53,** 239 (1954).
13. D. P. HACKETT, H. A. SCHNEIDERMANN and K. V. THIMANN, *Arch. Biochem.* **47,** 205 (1953).

CHAPTER 3

ENZYME-SUBSTRATE SPECIFICITY IN POTATO POLYPHENOL OXIDASE

J. LAVOLLAY, G. LEGRAND, G. LEHONGRE and J. NEUMANN

Conservatoire National des Arts et Métiers, Paris

CERTAIN facts have suggested that the specificity of soluble vegetable polyphenol oxidases depends upon their origin. Catechol has often been used as a substrate for determining *o*-diphenolase activity, but this diphenol is frequently not the best substrate for this determination. Among the *o*-diphenolic substrates acted on by *Agaricus campestris* tyrosinase, catechol consumes oxygen with the greatest velocity, i.e. almost four times the velocity obtained with chlorogenic acid.[1,2] When used as a carrier in the oxidation of ascorbic acid by this enzyme, catechol caused the velocity to be three times greater than for chlorogenic acid.[3]

The sweet-potato *o*-diphenolase acting on catechol, homocatechol and chlorogenic acid caused rates of oxidation in the ratio of 1:3:6, respectively.[2] The affinities ($1/K_m$) of tobacco root and leaf tyrosinases for chlorogenic acid are greater than for catechol[4,5] but tobacco seed tyrosinase shows the greatest affinity for DOPA.[6]

Baruah and Swain[7] have listed a number of diphenols as possible substrates for potato phenolase. Chlorogenic acid and some flavonoids could be oxidized but none of the glycosides. Similar results were obtained by Roberts and Wood[8] using an acetone powder preparation of tea leaf.

In this laboratory, the initial velocity of oxygen consumption has been determined using low concentrations of soluble potato tyrosinase with a large excess of ascorbic acid together with various *o*-diphenolic substrates, acting as carriers in low concentrations. It has been generally accepted that the overall velocity, under these conditions, depends only upon velocity of the *o*-diphenol oxidation, the ascorbic acid maintaining a constant concentration of the phenolic substrate for a duration of time sufficient to allow a precise measure of initial velocity, even with small concentrations of the substrate. Ascorbic acid also prevents inhibitions arising from the oxidation products. In our work it has been effectively verified that the initial velocities were unchanged when the molecular ratios of the ascorbic acid to the phenolic carrier were varied between 50 and 200. This result does not mean that ascorbic acid has no action upon the oxidizing system, but it suggests that within the range of concentrations used, the results do not depend upon the concentration of ascorbic acid.

With many phenolics, the concentration of the diphenol carrier appeared to remain unchanged as long as reduced ascorbic acid was present. But, in certain cases, namely with catechin or with caffeic acid, coloured products were formed long before the ascorbic acid was completely oxidized. Since

TABLE 1

Phenol	Medium without ethanol			Medium with ethanol		
	K_m (mM)	$\frac{V_m}{2}$ (mg atoms O_2/l./min)	$\frac{V_m}{2K_m}$	K_m (mM)	$\frac{V_m}{2}$ (mg atoms O_2/l./min)	$\frac{V_m}{2K_m}$
Quercetin (flavonol)				0·01(3)	0·19	14·6
Rhamnetin (flavonol)				0·03	0·22	7·3
Fisetin (flavonol)				0·04	0·19	4·7
Chlorogenic acid	0·01(7)	0·17	10·0	0·02(2)	0·08	3·6
Luteolin (flavone)				0·05	0·17	3·4
Homocatechol	0·06	0·21	3·5	0·11	0·22	2·0
3,4-Dihydroxyphenylpropionic acid	0·19	0·20	1·0	0·12	0·20	1·7
3,4-Dihydroxycinnamic acid	0·08	0·15	1·9	0·25	0·23	0·9
Catechol	0·11	0·11	1·0	0·22	0·12	0·5
Rutin (flavonol heteroside)				0·30	0·16	0·5
Taxifolin (flavanonol)	0·35	0·13	0·4	0·55	0·15	0·3
Catechin	0·65	0·25	0·4	0·74	0·14	0·2
3,4-Dihydroxyphenylacetic acid	0·55	0·18	0·3	1·50	0·30	0·2
Maclurin				0·24	0·03	0·1(2)
Quercitrin (flavonol heteroside)				0·36	0·05	0·1(4)
3,4-Dihydroxybenzoic acid	2·20	0·07	0·0(3)	0·50	0·01	0·0(2)

the coloured products appear to arise from the quinones, this result would indicate that in these cases, the quinone reduction was a slower reaction than the phenol oxidation.

Initial velocities were determined using the manometric method, and measured in terms of gram atoms of oxygen absorbed per minute at 30°. Each Warburg flask contained ascorbic acid (10^{-5} mole), the phenolic substrate solution, EDTA (sodium salt; 10^{-8} mole)* and EDTA-purified 0·1 M-phosphate buffer (2·5 ml; pH 5·7). The total solution (2·75 ml) was

* This concentration of EDTA inhibits ascorbic acid autoxidation but does not affect enzyme activity.

placed in the flask and the enzyme solution (0·25 ml) was put in the side arm.

The partially purified enzyme solution was obtained from an aqueous extract of potato peelings (var. *Bintje*), by fractionation with acetone and ammonium sulphate followed by dialysis. Alcoholic solutions were used for the water insoluble flavonoids. In these cases, the medium contained a final concentration of 5 per cent ethyl alcohol. As shown in Table 1, nine phenolic substrates were compared in media with and without alcohol. The V_m and K_m values were both affected by the solvent, but the values of $V_m/2K_m$ classify these nine substrates in the same order in alcoholic as in aqueous solutions.

With all the phenolic compounds listed in Table 1, straight lines have been obtained for variations of $1/v$ when plotted against $1/s$, following Lineweaver and Burk. Thus, the particular values for V_m and K_m could be obtained graphically.

It is apparent, from Table 1, that the phenolic substrates investigated can be placed in one order when V_m is taken as the basis of the comparison and a different order when $1/K_m$ is used. We have suggested[9] that the activities of the oxidizing systems (enzyme plus diphenolic carrier) should be best determined from the velocities of the reaction at equivalent relative concentrations for enzyme and phenol. Thus, the ratio $V_m/2K_m$ (Table 1) can be considered as a molecular velocity constant, characterizing the enzyme–substrate combination and allowing a classification of the different carriers, according to the activity of each so-called enzyme–substrate complex. It will be observed that this constant is half the value of the coefficient of the $1/s$ term in the Lineweaver–Burk expression of the Michaelis equation and gives half the slope of the straight line:

$$\frac{1}{v} = \frac{K_m}{V_m}\left(\frac{1}{s}\right) + \frac{1}{V_m}$$

In the table, the substrates have been placed in order of decreasing values for $V_m/2K_m$. This order indicates that potato tyrosinase shows a certain degree of adaptation to the $C_{(6)}$—$C_{(3)}$—$C_{(6)}$ structure which characterizes flavonoids and chlorogenic acid. The quercetin glycosides show a small affinity for the enzyme, but the aglycones (quercetin, rhamnetin, fisetin and luteolin) appear to be very good substrates. For natural conditions, water solubility appears to be an important property for a carrier, and thus the specificity of potato tyrosinase for the soluble chlorogenic acid presents a special significance.

REFERENCES

1. D. Keilin and T. Mann, *Proc. Roy. Soc.* **B125,** 187 (1938).
2. I. Z. Eiger and C. R. Dawson, *Arch. Biochem.* **21,** 194 (1949).

3. J. LAVOLLAY, G. LEGRAND, G. LEHONGRE and J. NEUMANN, *Qualitas Plant. Mat. Veget.* **3–4,** 521 (1958).
4. E. C. SISLER and N. J. EVANS, *Plant Physiol.* **33,** 255 (1958).
5. R. A. CLAYTON, *Arch. Biochem.* **81,** 404 (1959).
6. A. FREY-WYSSLING and J. W. SZARKOWSKI, *Experientia* **15,** 462 (1959).
7. P. E. BARUAH and T. SWAIN, *J. Sci. Food Agric.* **10,** 125 (1959).
8. E. A. H. ROBERTS and D. J. WOOD, *Nature, Lond.* **167,** 608 (1951).
9. J. NEUMANN, G. LEGRAND, G. LEHONGRE and J. LAVOLLAY, *Compt. Rend.* **251,** 3091 (1960).

CHAPTER 4

SOME PHYSIOLOGICALLY IMPORTANT AROMATIC HYDROXYLATION SYSTEMS

J. B. JEPSON

Courtauld Institute of Biochemistry, Middlesex Hospital Medical School

IT IS reasonably certain that many of the phenolic constituents of plants are formed by a series of enzymic reactions, starting either from glucose via shikimic and related acids and leading to phenols with hydroxyls on adjacent carbon atoms,[1] or from linear condensation of acetyl- and malonyl-coenzyme-A units followed by ring closures and leading, therefore, to phenols with hydroxyls on alternate carbon atoms.[2] Of the enzymes responsible for these biosyntheses we know practically nothing. It is also certain that these processes for phenol formation play no part in animal metabolism, for few of the higher animals are able to conduct the synthesis of the isolated aromatic ring. (Introducing the word *animal* into the proceedings of the Plant Phenolics Group recalls the comment by the neurochemist Heinrich Waelsch that if laboratory workers representing each of the separate biological disciplines were lined up in front of a collection of animal tissues, you could always tell the neurochemist because he would be the one rushing for the piece of liver. Similarly, it is wise, in the interests of plants, to be alert to the biochemical potentialities of human liver.)

Animal tissues, and particularly liver, possess many hydroxylation systems capable of the enzymic substitution of hydroxyl groups for hydrogen at aliphatic or aromatic carbon atoms, or for the addition of hydroxyls to unsaturated carbons. The pathways of steroid synthesis starting with the oxidative ring closure of squalene[3] and the pathways of "detoxicative" metabolism[4] provide many examples. The systems upon which most experimental work has been expended are those catalysing direct hydroxylation into preformed aromatic rings, and this survey is concerned only with such systems. To narrow the field still further, only three aromatic hydroxylation reactions will be considered in detail, namely: (i) the hydroxylation of the amino-acid phenylalanine in the *para*-position to give the phenolic amino-acid tyrosine; (ii) the hydroxylation of the amino-acid tryptophan in the 5-position to give 5-hydroxytryptophan; and (iii) the hydroxylation of indoles other than tryptophan in the 6-position. The choice has not fallen on these three because of their simplicity (for indeed their enzymic mechanisms have not yet been elucidated), nor because of their similarity (because, in fact, they differ greatly from one another and from any model system yet devised), but because each gives, as end product, a stable, monohydric phenol of undoubted or plausible physiological significance.

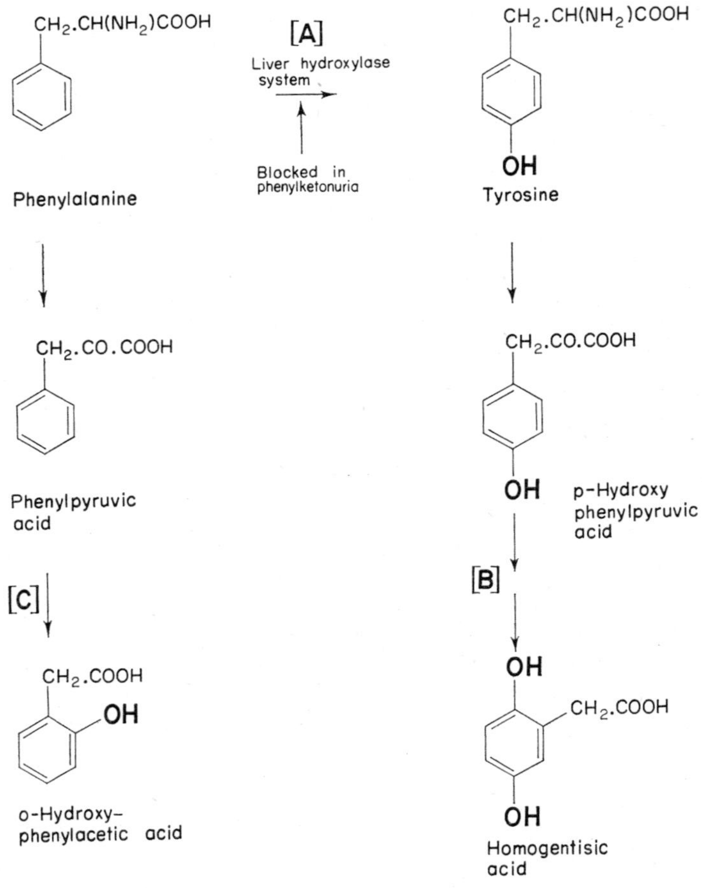

FIG. 1. Hydroxylative metabolism of phenylalanine in animals.

The hydroxylation of phenylalanine to tyrosine is a metabolic reaction common to all animals. It can be considered the first step of the metabolic sequence for the dissimilation or utilization of aromatic amino-acids (Fig. 1, reaction A). Subsequent steps involve other types of hydroxylation, for example the enzymic hydroxylation of *p*-hydroxyphenylpyruvic acid with a simultaneous 1,2 shift of the side chain, and which requires ascorbic acid as a cofactor to maintain accessory quinones in a reduced form[5]

(Fig. 1, reaction B). The enzyme system for tyrosine formation, phenylalanine *p*-hydroxylase, has been found solely in liver,[6] in the soluble supernatant after extended centrifugation of disintegrated liver tissue. The oxygen of the introduced hydroxyl group does not come from water but from free molecular oxygen which is obligatory for the reaction. The enzyme requires a soluble cofactor, known to be a light-sensitive unconjugated pteridine at the dihydro level of reduction.[7] As hydroxylation proceeds, the cofactor is oxidized and must be regenerated by reduction with reduced nicotinamide adenine dinucleotides, $NADH_2$ or $NADPH_2$, a separate liver enzyme, pteridine reductase, being required for this regeneration.

Only the L-enantiomorph of phenylalanine is hydroxylated and no hydroxylation at the *ortho*- or *meta*-positions can be detected. Such specificity eliminates a hydroxylation mechanism dependent solely on the production of a general hydroxylating agent like hydrogen peroxide. However, the hydroxylation is not absolutely specific for L-phenylalanine. For example, *p*-fluorophenylalanine also gives tyrosine, the fluorine (instead of the hydrogen) being replaced by hydroxyl.[8] An interesting parallel is the metabolic replacement of chlorine by hydroxyl when *Aspergillus niger* is grown on *o*-chlorophenoxyacetic acid as sole source of carbon, although in this case every possible substitution of hydroxyl into the ring occurs as well.[9]

A great deal of study is being devoted to phenylalanine hydroxylase because of its medical importance. Its presence in human liver is genetically controlled, some people lacking the genetic information necessary for its production. People who are homozygous for lack of the enzyme (i.e. neither of the chromosomes of the appropriate chromosome pair carry the gene for hydroxylase production) are unable to convert dietary phenylalanine to tyrosine. They have a complete enzyme block at this point and phenylalanine metabolism has to be diverted elsewhere, largely to the ketone, phenylpyruvic acid, and thence to *o*-hydroxyphenylacetic acid (Fig. 1, reaction C). The latter *ortho*-hydroxylation[10] (site and mechanism unknown) further exemplifies the extraordinary diversity and specificity of metabolic hydroxylation. This metabolic abnormality, phenylketonuria, is a serious one, leading to patients who are adequate physically but grossly retarded mentally. Heterozygotes for the abnormality, with only one gene for hydroxylase production, have sufficient enzyme to cope with the usual dietary amounts of phenylalanine but their lower-than-normal metabolic potentialities can be shown by challenging them with a large dose of phenylalanine, when their blood tyrosine concentration only shows half the rise shown by normals (full phenylketonurics, of course, show no rise in blood tyrosine).[11]

The finding that *para*- and *meta*-hydroxylation processes can be associated may provide a clue to the mechanism of phenylalanine hydroxylase.

Normal human urine contains all three of the isomeric tyramines: considerable amounts of *p*-tyramine, largely from tyrosine; very small amounts of *m*-tyramine; and *o*-tyramine from unknown sources. Phenylketonurics, as might be expected, excrete *less* *p*-tyramine than normals; they also excrete less *m*-tyramine, though the excretion of *o*-tyramine is normal.[12] So it is possible that *para*- and *meta*-substitutions are linked at some step of the hydroxylation, say as a 3,4-epoxide, cleavage of which would give largely a *p*-hydroxy derivative with very much smaller amounts of *meta*-isomer, but both products would be diminished if the enzyme for epoxide formation was deficient. Work on these possibilities is in hand, but the first requirement is to find whether *m*-tyrosine or phenylethylamine or some other compound is the precursor of *m*-tyramine.

The second reaction for consideration, the specific hydroxylation of L-tryptophan to 5-hydroxy-L-tryptophan (5-HTP), is superficially similar to the phenylalanine hydroxylase reaction, and indeed it has recently been shown that the specificity of the latter reaction has to be extended to include L-tryptophan, with 5-hydroxy-L-tryptophan as the only detectable product.[13,14] The rate of hydroxylation of tryptophan is only one-thirtieth of that of phenylalanine, but all requirements for enzymes, cofactors, carriers and oxygen are identical for both systems. Nevertheless, it seems certain that this enzyme system is *not* the one responsible for the large-scale hydroxylation of tryptophan that undoubtedly occurs in the human and animal body. For one thing, phenylalanine hydroxylase is only known as a liver system, while 5-hydroxytryptophan and its metabolites are never found in liver but in gut and brain, tissues which have never been shown to hydroxylate phenylalanine. Many whole organisms produce 5-hydroxytryptophan, the precursor of the physiologically-important amine 5-hydroxytryptamine (5-HT, serotonin) (Fig. 2, reaction A). For example, the violet pigment of *Chromobacterium violaceum*, violacin, is derived from 5-hydroxytryptophan, the precursor of which is free tryptophan.[15] Tryptophan hydroxylation in man can be shown by inhibiting the general aromatic amino-acid decarboxylase system[16] which usually produces the hydroxyamine from the hydroxyamino-acid too rapidly for the latter to be demonstrated. The effect is best seen in cases of carcinoid syndrome due to a tumour of the gastric enterochromaffin cells, where administration of the decarboxylase inhibitor α-methyl-dopa causes large amounts of 5-hydroxytryptophan to be excreted in the urine.[17] The physiological site of tryptophan hydroxylation has not yet been found, but intestinal mucosa must be regarded as a strong candidate in view of its high concentration of serotonin which plays some part in maintaining intestinal motility. A claim has been made,[18] but not substantiated,[13] that a particulate preparation from gut can catalyse this hydroxylation when supplemented with cupric ions and ascorbic acid acting anaerobically; probably this preparation causes some non-enzymic (and non-physiological) hydroxylation, similar to that performed by the well-known ferrous chelate–ascorbate–oxygen system.[19]

Serotonin is certainly indispensible as a neurotransmitter substance in some lowly forms of life, but some ingenious roles are being attributed to

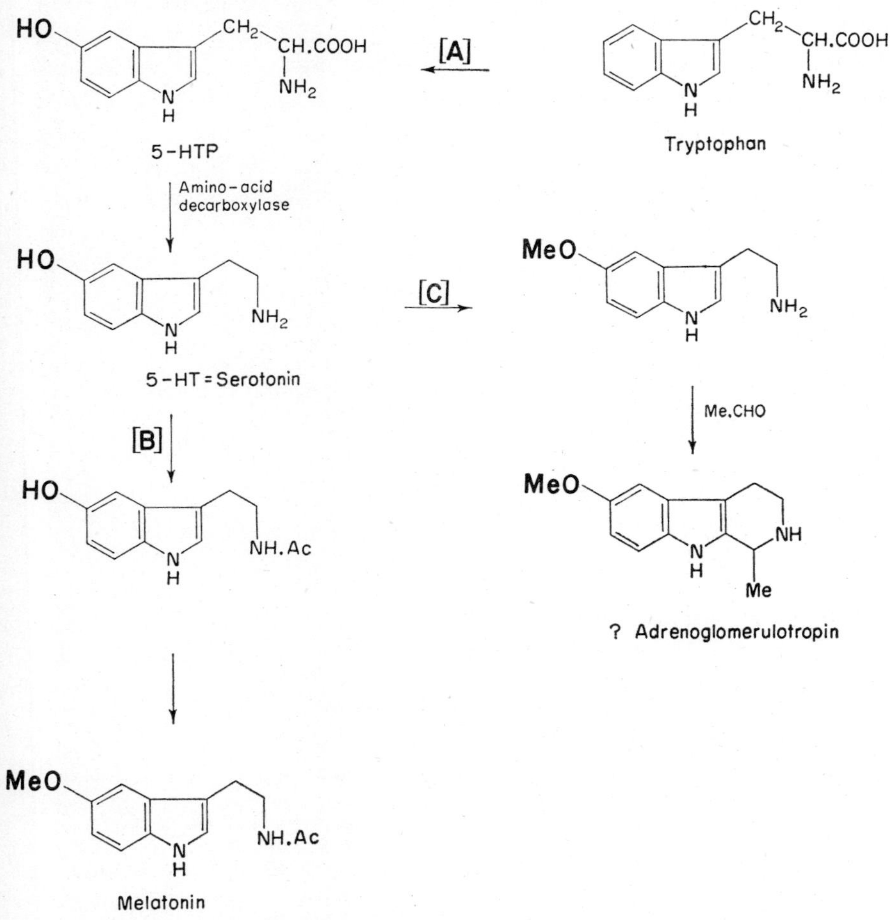

FIG. 2. Hydroxylative metabolism of tryptophan in animals.

its derivatives in mammals including man. Attention has been focused on the high serotonin content of the pineal gland, an organ considered to be a vestigial remnant from a remote non-mammalian past. From the pineal can be isolated *melatonin*[20] (*O*-methyl-*N*-acetylserotonin) and the enzymes which catalyse its synthesis from serotonin[21,22] (Fig. 2, reaction B). Melatonin acts on melanocyte cells causing clumping of the melanin granules contained therein so that it promotes the lightening of dark skin.

Following another metabolic pathway (Fig. 2, reaction C) from serotonin, it has been claimed that the pineal manufactures a tetrahydro-β-carboline hormone[23] (*adrenoglomerulotropin*) with the property of stimulating aldosterone secretion and so controlling Na^+ retention.[24] Both these regulatory functions, for skin colour and for ionic balance, can obviously be regarded as developing from a semiaqueous reptilian ancestry.

The third reaction nominated for consideration was discovered following the work of Brodie on the metabolic potentialities of the microsome fraction of animal liver.[25] Liver microsomes were found capable of handling a wide variety of lipid-soluble materials, indigenous or foreign to the cell, by oxidative processes including processes of aromatic hydroxylation.[26] All these reactions were enzymic, with an absolute requirement for molecular oxygen and for reduced – NADP (TPNH); the other nicotinamide dinucleotides were no substitute (contrast phenylalanine

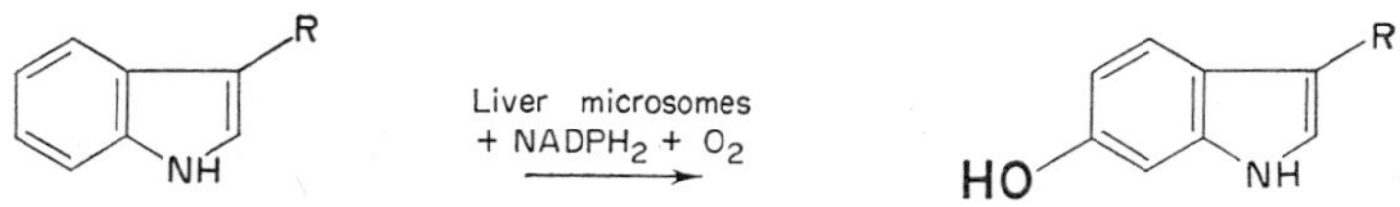

FIG. 3. Microsomal hydroxylation of indoles.
R = $—CH_2CH_2NH_2$
$—CH_2CH_2NMe_2$
$—CH_2CH(NHAc)COOH$
$—CH_2COOH$
$—CH_3$
etc.
No oxidation when R = $—CH_2CH(NH_3^+)COO^-$
Oxidation at C—3 when R = H.

hydroxylase). For example, aniline was hydroxylated almost exclusively in the *para*-position by rabbit liver microsomes, but in the *ortho*- and *para*-positions equally by cat liver microsomes; this paralleled exactly the findings when rabbit and cat were dosed with aniline and the urines examined for aminophenol metabolites.

When the rabbit liver microsomal system is applied to tryptamine the sole product is 6-hydroxytryptamine in high yield, with no trace of the (originally expected) 5- and 7-hydroxytryptamines[27] (Fig. 3). The same general reaction holds for a wide range of indoles, whether basic, acidic or neutral, provided they are lipid-soluble. *In vitro* these hydroxylations, requiring oxygen and $NADPH_2$, are catalysed by the microsomal fraction from animal liver but by no other tissue. In the whole animal too, including man, 6-hydroxylation is a major metabolic pathway for fat-soluble indoles, provided other preferred alternative routes are blocked or swamped.

Tryptophan, lipid-insoluble under all circumstances, is *not* hydroxylated by microsomes. Lipid-soluble *N*-acetyltryptophan is hydroxylated, both enantiomorphs reacting at the same rate (Jepson, unpublished work), so either the enzymic reaction is indifferent to the stereo configuration of its substrate (an unlikely event) or the hydroxylation is not the primary enzymic step. The plant auxin indolylacetic acid is 6-hydroxylated, but there is no evidence for 6-hydroxyindolylacetic acid occurring in plants. *N,N*-Dimethyltryptamine gives a high yield of 6-hydroxy-derivative using isolated microsomes or in the whole animal. Humans are especially effective in this conversion, which raises a point of great pharmacological and clinical interest, because the product, 6-hydroxy-*N,N*-dimethyltryptamine, is a positional isomer of the natural plant indoles bufotenin (5-isomer) and psilocin (4-isomer) and, like them, a highly potent hallucinogen, causing gross visual and aural aberration and behavioural derangement. Thus the metabolic machinery of a human receiving dimethyl tryptamine seeks to rid itself of this foreign material by a quite general hydroxylation reaction, but thereby converts it into a far more dangerous derivative.[28] It also seems possible that 6-hydroxyindoles produced in the body could ring-close under physiological conditions *in vivo* to give β-carbolines like harmalol, a potent hypotensive agent.

It is not yet certain which of the cell organelles and fragments that are brought together during the preparation of the "microsome" fraction carry the hydroxylation system, but it seems likely that portions of the membranes of the smooth-surfaced endoplasmic reticulum are responsible.[29,30] Several different lines of investigation have hinted at the existence of a *family* of $NADPH_2$-requiring aromatic hydroxylases in liver microsomes, with different specificities, pH optima and species distribution.[31,32] Nevertheless, a single common enzyme could still be the responsible agent, but dependent on the transport properties of the surrounding microsomal membranes for the supply of substrates.[33] Special disruptive techniques have recently enabled the hydroxylating activity of liver microsomes for acetanilide to be solubilized, but then a special lipid cofactor has to be added separately.[34,35]

There are plenty of theories for the mechanism of microsomal hydroxylation, but too many have been based on dubious analogies with non-enzymic hydroxylating systems like the ferrous chelate–ascorbate–oxygen system.[36–38] It is certain that the oxygen atom of the hydroxyl group comes from molecular oxygen and not from water[39] (this is not always the case with enzymic hydroxylations which at first sight resemble microsomal hydroxylation, e.g. the 6-hydroxylation of nicontinic acid [40]). It is also certain that no hydrogen from the aqueous medium enters a stable position in the hydroxylated product.[39]

Whatever the mechanisms of enzymic hydroxylations,[41] there is no doubt that they differ significantly from the more random hydroxylations brought about by hydrogen peroxide or by many of the non-enzymic systems generating free radicals.[42] However, high specificity of substituent position does not exclude the hydroxylating agent being a free radical. Norman has shown[43] how the electrophilic character of the hydroxyl radical can allow specific non-random substitution even in homogeneous solution, to say nothing of the situation when both substrate and oxidizing system are subjected to the circumscribing environment of an enzyme.

REFERENCES

1. B. D. DAVIS, in *A Symposium on Aminoacid Metabolism*, p. 799, eds. W. D. McElroy and H. B. Glass, University of Baltimore (1955).
2. A. J. BIRCH, *Proc. Chem. Soc.* 3 (1962).
3. J. W. CORNFORTH, *J. Lipid Res.* **1,** 3 (1960).
4. R. T. WILLIAMS, *Detoxication Mechanisms*, 2nd Edn., Chapman Hall, London (1959).
5. V. G. ZANNONI, *J. Biol. Chem.* **237,** 1172 (1962).
6. C. MITOMA, *Arch. Biochem. Biophys.* **60,** 476 (1956).
7. S. KAUFMAN, *J. Biol. Chem.* **237,** PC2712 (1962).
8. S. KAUFMAN, *Biochem. Biophys. Acta* **51,** 619 (1961).
9. J. K. FAULKNER and D. WOODCOCK, *J. Chem. Soc.* 5397 (1961).
10. R. E. TASHIAN, *Science*, **129,** 1553 (1959).
11. G. A. JERVIS, *Clin. Chim. Acta* **5,** 471 (1960).
12. J. B. JEPSON, W. LOVENBERG, P. ZALTZMAN, J. A. OATES, A. SJOERDSMA and S. UDENFRIEND, *Biochem. J.* **74,** 5P (1960).
13. J. RENSON, H. WEISSBACH and S. UDENFRIEND, *J. Biol. Chem.* **237,** 2261 (1962).
14. R. A. FREEDLAND, I. M. WADZINSKI and H. A. WAISMAN, *Biochem. Biophys. Res. Comms.* **5,** 94 (1961).
15. C. MITOMA, H. WEISSBACH and S. UDENFRIEND, *Arch. Biochem. Biophys.* **63,** 122 (1956).
16. W. LOVENBERG, H. WEISSBACH and S. UDENFRIEND, *J. Biol. Chem.* **237,** 89 (1962).
17. J. A. OATES and A. SJOERDSMA, *Amer. J. Med.* **32,** 333 (1962).
18. J. R. COOPER and I. MELCER, *J. Pharm. Exp. Therap.* **132,** 265 (1961).
19. C. E. DALGLEISH, *Arch. Biochem. Biophys.* **58,** 214 (1955).
20. A. B. LERNER, J. D. CASE and Y. TAKAHASHI, *J. Biol. Chem.* **235,** 1992 (1960).
21. H. WEISSBACH, B. G. REDFIELD and J. AXELROD, *Biochem. Biophys. Acta* **54,** 190 (1961).
22. J. AXELROD and H. WEISSBACH, *J. Biol. Chem.* **236,** 211 (1961).
23. W. M. McISAAC and G. FARRELL, *Arch. Biochem. Biophys.* **94,** 543 (1961).
24. G. FARRELL, W. M. McISAAC and A. N. TAYLOR, *Excerpta Medica Intern. Congress Series* No. **51,** Abstract 14 (1962).
25. B. B. BRODIE, *J. Pharm. Pharmacol.* **8,** 1 (1956).
26. C. MITOMA, H. S. POSNER, H. C. REITZ and S. UDENFRIEND, *Arch. Biochem. Biophys.* **61,** 431 (1956).
27. J. B. JEPSON, P. ZALTZMAN and S. UDENFRIEND, *Biochem. Biosphys. Acta.* **62,** 91 (1962).
28. S. SZARA, *Experientia* **17,** 76 (1961).
29. J. ROTHSCHILD, *Biochem. J.* **83,** 17P (1962).
30. J. FOUTS, *Biochem. Biophys. Res. Comms.* **6,** 373 (1961).
31. H. S. POSNER, C. MITOMA and S. UDENFRIEND, *Arch. Biochem. Biophys.* **94,** 269 (1961).
32. P. J. CREAVEN, D. V. PARKE and R. T. WILLIAMS, *Biochem. J.* **85,** 5p (1962).
33. B. B. BRODIE and C. A. M. HOGBEN, *J. Pharm. Pharmacol* **9,** 365 (1957).
34. Y. IMAI and R. SATO, *Biochem. Biophys. Acta* **42,** 164 (1960).
35. K. KRISCH and H. STAUDINGER, *Biochem. Biophys. Res. Comms.* **4,** 118 (1961).
36. R. M. ACHESON and C. M. HAZELWOOD, *Biochem. Biophys. Acta* **42,** 49 (1960).

37. R. Breslow and L. N. Lukens, *J. Biol. Chem.* **235,** 292 (1960).
38. A. Hanaki and M. Ishidate, *Biochem. Biophys. Acta* **57,** 180 (1962).
39. H. S. Posner, C. Mitoma, S. Rothberg and S. Udenfriend, *Arch. Biochem. Biophys.* **94,** 280 (1961).
40. A. L. Hunt, D. E. Hughes and J. M. Lowenstein, *Biochem. J.* **66,** 2P (1957).
41. L. Massart and R. Vercauteren, *Ann. Rev. Biochem.* **28,** 527 (1959).
42. A. Weissler, *Nature, Lond.* **193,** 1070 (1962).
43. R. O. C. Norman and G. K. Radda, *Proc. Chem. Soc.* 138 (1962).

CHAPTER 5

TOCOPHEROLS

J. GREEN

Walton Oaks Experimental Station, Tadworth, Surrey

INTRODUCTION AND CHEMISTRY

ALTHOUGH the tocopherols, in their own right, undoubtedly constitute an interesting and important group of natural phenols, whose chemistry alone would provide the basis for a lengthy lecture, they have a special significance to chemists and biochemists because of their identification with vitamin E. They thus enter the extensive realms of animal nutrition and metabolism. The author of a talk on the tocopherols, therefore, must, above all things, decide where to begin and when to stop. It would seem impossible to deal, on the one hand, with those aspects of the subject most likely to interest members of this symposium (who are mainly concerned with plant chemistry and biochemistry) and, on the other hand, with the vast and often contradictory literature on vitamin E itself. It is clear that the subject must be limited, and I have in fact done this quite simply, if rather severely, by deciding to eliminate practically all mention of the animal kingdom, except where it is especially relevant. This has the advantage of immediately disposing of the necessity of discussing vitamin E as a vitamin at all (and of about 99·9 per cent of the available literature). It does enable me to deal at much greater length than would otherwise be possible with the tocopherols purely as a group of plant phenols and, in sympathy with the general aims of this meeting, to elaborate on their chemistry and plant metabolism. Actually, in spite of the immense amount of work that has been done on vitamin E in animals, very little is known of the relationships of tocopherols to plant metabolism and what there is seems to be of comparatively recent origin. It seems possible, however, that recent discoveries in the field of plant quinones will lead to greater interest in the role of tocopherols in plants, especially the higher plants.

It is just over forty years ago that Evans and Bishop[1] reported that female rats failed to reproduce on certain purified diets and recognized that a factor was missing, which they eventually termed vitamin E. Over a decade later, this factor was identified with a highly-reducing lipid-like phenolic substance, which was called α-tocopherol. Soon afterwards, two other substances were discovered, obviously closely similar to α-tocopherol but less active in restoring fertility to the female rat, and these were called β- and γ-tocopherol. In the ensuing years, vitamin E was found to be an essential nutrient for a great variety of animal species. The structure of the three known tocopherols was discovered and the substances were synthesized. It was confirmed that α-tocopherol was the most active substance and it is now known to be preferentially metabolized in all species so far examined. So much so, that examination of many animal tissues reveals it as the only tocopherol present in the majority of cases, even when the diet of the animals has contained appreciable quantities of other tocopherols. In recent years, more members of the tocopherol family have been discovered, but none of them equals α-tocopherol in

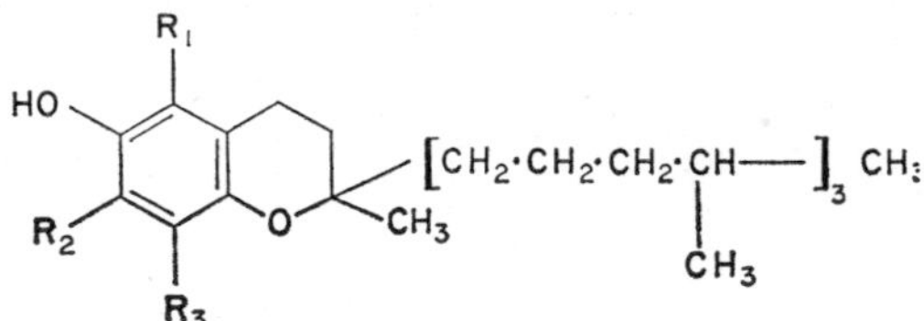

(I)

α-Tocopherol: $R_1 = R_2 = R_3 = CH_3$

β-Tocopherol: $R_1 = R_3 = CH_3$; $R_2 = H$

γ-Tocopherol: $R_2 = R_3 = CH_3$; $R_1 = H$

δ-Tocopherol: $R_3 = CH_3$; $R_1 = R_2 = H$

η-Tocopherol: $R_2 = CH_3$; $R_1 = R_3 = H$

ζ_2-Tocopherol: $R_1 = R_2 = CH_3$; $CH_3 = H$

potency. A bewildering array of deficiency dis eases is now known to occur in various species on deficient diets. Thus, the male rat becomes reversibly sterile and the female rat irreversibly so; the rabbit suffers a severe and often fatal muscular dystrophy that is reversible with vitamin E; other species, such as the sheep, appear to suffer a dystrophy that is irreversible; chickens suffer from vascular disease and brain damage; mice and pigs suffer a severe necrotic degeneration of the liver, which can be fatal; and the elegant mink and the humble kitten both suffer a disease of adipose tissue, generally called "yellow fat" disease. In spite of the abundance of experimental material, however, the role of tocopherols in animals cannot at present be adequately understood. One dearly wishes, indeed, that there were less diseases and more biochemistry. It is not yet clear whether the α-tocopherol can take part in oxidation-reduction systems or is metabolized by animals to an intermediate with a more fundamental role. Much of the background for suggested functions of the vitamin, however, comes from a study of its chemistry and, at this point, we leave the animal kingdom and consider the tocopherols themselves.

Chemically, the tocopherol fraction of plant lipids is now known to comprise two series of compounds, all derivatives of chroman-6-ol. The first series (I) is derived from a 2,2-disubstituted compound, tocol, which bears a saturated $C_{(16)}$ side chain; the second series (II), discovered more recently,[2] is derived from a compound, tocotrienol, which bears an unsaturated isoprenoid $C_{(16)}$ side chain at $C_{(2)}$. The second series can thus

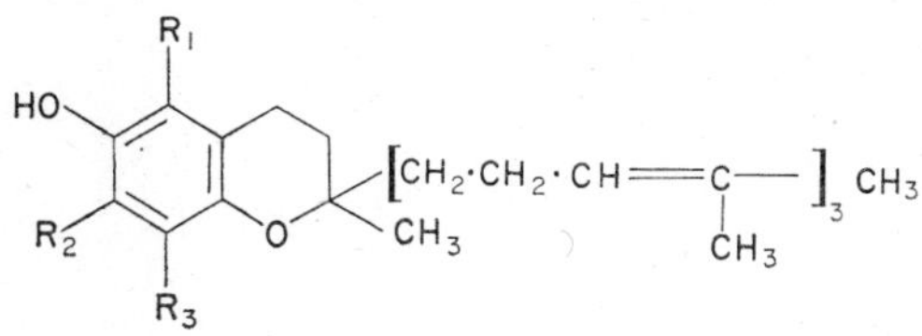

(II)

ε-Tocopherol: $R_1 = R_2 = CH_3$; $R_2 = H$

ζ_1-Tocopherol: $R_1 = R_2 = R_3 = CH_3$

be readily converted to the first series by side chain hydrogenation. The tocopherols are derived from these two compounds by nuclear methylation. There are thus, including in each case the unmethylated starting compound, eight possible derivatives in each series. In the tocol series, six have so far been found in plants, giving the series, in order of discovery, α-, β-, γ-, δ-, η-, and ζ_2-tocopherol. Neither tocol itself nor 5-methyltocol have so far been observed in natural products, although the finding of the latter would be difficult in any case as it can only be distinguished from other tocopherols by special analytical procedures.[3] Of the second series only two members have so far been found, ε-tocopherol and ζ_1-tocopherol (the 5,8-dimethyl and 5,7,8-dimethyltocotrienols respectively); but they are of especial importance as they form an appreciable part of the tocopherol fraction of the staple cereals, such as wheat and rye.

The tocopherols are pseudo-phenols and can be readily extracted from Claisen's alkali. The strength of the phenolic function is decreased by increase in nuclear methylation, and δ-tocopherol and tocol are strong enough phenols to form solid sodio-derivatives.[4] The substances are all oils except ζ_2-tocopherol, which is a solid at room temperature. *d*-α-Tocopherol, when highly pure, melts at about 3°, but the *dl* form has not been obtained solid. The tocopherols are all strongly reducing to reagents such as ferric chloride and silver nitrate and are fat antioxidants, whose activity depends largely on the degree of nuclear methylation. They are reasonably stable in air when pure, can be heated to 250° in the absence of air for prolonged periods without change, are stable to strong acids, but rapidly oxidized by alkalis in the presence of oxygen. All members of the tocol series have now synthesized, even 5-methyltocol, but there are difficulties in the synthesis of the tocotrienol series.[5] Most of the known chemistry of the tocopherols depends on the readiness with which they can be oxidized. The oxidation-reduction reactions have only been worked out in detail for α-tocopherol.[6] It can be oxidized by a number of mild oxidizing agents to yield the corresponding *p*-quinone (III), which in turn can be reduced to a quinol, or alternatively reductively re-cyclized to α-tocopherol. In the presence of 2,2′-dipyridyl, however, ferric chloride oxidation in ethanol stops at an intermediate oxidation product (IV),

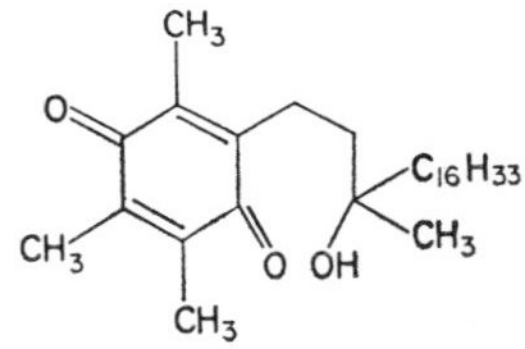

III

once called tocopheroxide, now known to be a mono-acetal of the *p*-quinone, tocopherethoxide.[7] It may be formed through an intermediate carbonium ion by loss of electrons and a hydrogen ion, followed by the reversible addition of a hydroxyl ion. Although the substance does not occur in biological systems, a hypothetical hemiketal could provide an

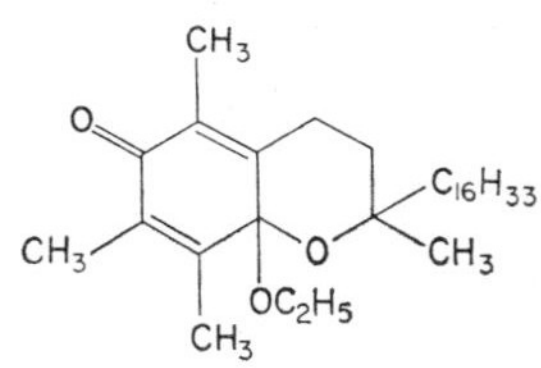

IV

intermediate step in the biological oxidation of tocopherol to the *p*-quinone. Tocopherethoxide is readily reduced by ascorbic acid to α-tocopherol and thus has appreciable biological activity: much more so than the *p*-quinol, which must cyclize to give α-tocopherol. Whether this step is not readily carried out *in vivo* or whether solubility relationships are paramount is not known, but either of these may account for the weak biological activity of the *p*-quinol.[8] Dehydration of the *p*-quinone to either an unsaturated allylic quinone (cf. the ubiquinones) or a chromenol by loss of the tertiary hydroxyl group cannot be carried out, cyclization being the preferred reaction.[9] Little is known about the detailed oxidation of the other tocopherols. Eggitt and Norris[10] compared the ferric chloride–dipyridyl oxidations of five tocopherols and showed that the course of reaction was markedly affected by both the number and position of the nuclear methyl groups. Thus δ-tocopherol gave little tocopherethoxide or *p*-quinone. The final oxidation product of α-tocopherol is a red *o*-quinone (V), produced by nitric acid and also by prolonged $AuCl_3$ oxidation,[9] the 5-methyl group being eliminated during the reaction. The reaction is given by other tocopherols, all of which can, under certain conditions, give *o*-quinones;

but Eggitt and Norris(10) have again shown that nuclear methylation exerts a stabilizing effect. The oxidation of α-tocopherol with nitric acid in ethanol yields not *o*-quinone, but a variety of other products. α-Tocopherylquinone undergoes an interesting dismutation reaction in the presence of methanol and weak acids, forming α-tocopherol (presumably taking hydrogen atoms from the solvent) and also losing the 5-methyl group oxidatively to give *o*-quinone. Both reactions involve cyclization.

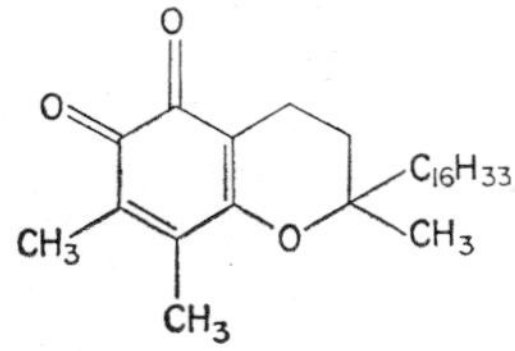

V

A further interesting difference exists in the chemistry of the methylated tocols, in that there is a pronounced non-equivalence in positions 5 and 7 of the aromatic ring. The origin of this effect, which cannot be discussed in any detail here, is related to (and is even more pronounced than) the Mills–Nixon effect in 6-tetralol but it leads to an enhancement of reactivity at $C_{(5)}$, compared to $C_{(7)}$. Thus all electrophilic reagents attack $C_{(5)}$ preferentially to $C_{(7)}$ if both positions are vacant. The lablity of the 5-methyl group under oxidative conditions is only one consequence of this difference. The presence or absence of a 5-methyl group markedly affects the antioxidant potency of the substance(11) and 5-methylated derivatives are more active as vitamin E in several biological systems than corresponding 7-methylated derivatives.

ANALYSIS AND DISTRIBUTION IN PLANTS

The analytical separation and determination of this group of substances is of especial importance because of their biological activity as vitamin E. Most vegetable oils, especially seed oils, contain appreciable quantities of tocopherols. Their vitamin E activity for animals is, however, exceedingly variable as it depends on which tocopherols are present and in what proportions. There is not only a very large difference in the biological potency of the different tocopherols (over a range of a hundredfold), but their distribution in the vegetable oils of commerce is varied. Since eight substances are involved, it is clear that the vitamin E potency of a vegetable material should be related to an individual determination of each substance. Since the biological activity of each of these eight substances is known, it is then possible to arrive at a reasonable estimate of the activity of the mixture.

The only way to do this is by some kind of chromatographic separation. Both adsorption(12) and partition(13) column chromatography provide a partial solution to the problem. Two-dimensional paper chromatography,[14] which we have used for several years, is a rapid and effective method of separation and when combined with the nitrosation technique of Quaife,[15] separates all the eight known tocopherols, which can then be determined by their reducing reaction.[16] By such methods Brown[17] has published analyses of a number of plants, including six grasses and various cereals. Green, Marcinkiewicz and Watt[14] have given the tocopherol contents of a number of cereal grains and several other seed oils. It is possible to discuss certain features in these results. First it is clear that tocopherol distribution falls into two rather distinct patterns. The more common one in higher plants is the $\alpha\gamma\delta$ pattern, found in most of the important seed oils, such as soybean, linseed oil, arachis oil, cotton-seed oil, etc.; less common, but of more importance for human nutrition, is the $\alpha\beta\epsilon\zeta_1$ pattern found in the staple cereals, wheat, barley and rye. Sometimes the pattern is incomplete (there is no δ-tocopherol in cotton-seed oil, for example) and occasionally minute traces of γ- and δ-tocopherol can be observed during some stages of the life-history of a plant normally containing only the $\alpha\beta\epsilon\zeta_1$ pattern, but, in general, the distinction seems valid. Whether there is any particular significance in the two patterns is unknown. The two patterns most likely reflect a different mode of biosynthesis in the two types of plant, but whether this in any way implies a different function for the resulting tocopherols cannot be answered. Rice presents an abnormal tocopherol pattern. It contains only α-tocopherol and the two rarest tocopherols, η-tocopherol and ζ_2-tocopherol. Ward has detected η-tocopherol in palm oil,[18] but no other source of ζ_2-tocopherol is known. Recently, Kubin and Fink[19] have analysed several species of mushroom and with the exception of *Boletus edulis,* which contained a little δ-tocopherol, they all contained the $\alpha\beta\zeta$ pattern, previously found only in cereals. Unfortunately these authors did not use two-dimensional chromatography and did not differentiate ϵ-tocopherol, nor were they able to distinguish ζ_1- from ζ_2-tocopherol.

The second feature that emerges from a study of these plant analyses is that although the pattern varies, almost without exception α-tocopherol is always present, although its proportion is extremely variable. The only lipid we have ever found with barely any α-tocopherol is kapok seed oil, which contains 0·64 mg/g of tocopherol, almost all γ-tocopherol. Apart from this it contained traces of α- and δ-tocopherol, too small to measure, This was, it is true, found in only one specimen, but it was fresh and carefully prepared. It is of interest that kapok seed oil is extensively used as an extender of cotton-seed oil, with which it is otherwise almost identical in composition. Cotton-seed oil, however, is a rich source of α-tocopherol.

The third feature that emerges is that whatever the pattern, in seeds it differs remarkably from that found in the leaves and stems of green plants, which generally contain only α-tocopherol, although sometimes this is accompanied by smaller amounts of other tocopherols. Out of this observation springs most of our present knowledge about the distribution and metabolism of tocopherols in plants.

TOCOPHEROL METABOLISM IN HIGHER PLANTS

The first comprehensive study of tocopherol metabolism in plants was carried out by the author[20] in 1958. Four plants were studied through one complete life-cycle. These were wheat and barley, whose seeds contain the $\alpha\beta\epsilon\zeta_1$ pattern, and pea and maize, which contain the $\alpha\gamma\delta$ pattern. The complete plant was, in each case, analysed for all the component tocopherols at several stages of the cycle. In each case, the seed started off with a depot of non-α-tocopherols. In the initial stages of germination and growth, there was found to be very little change in the total tocopherol content of the plant, but during the first few days there was a remarkable change in the components of the tocopherol fraction. The non-α-tocopherols disappear and their place is taken by α-tocopherol. By the time the first green shoots appear, net synthesis begins and thereafter an increase in α-tocopherol takes place throughout the plant. Very little, if any, synthesis of the other tocopherols takes place in wheat or barley whilst the plant is growing. As soon as the green, unripe ears appear, however, (in about three months growing time under our conditions) other tocopherols can be found. They appear, moreover, in a sequence that is the reverse of the order in which they originally disappear from the seed. In the main, they are only found in the developing seeds, the remainder of the plant continuing to contain α-tocopherol until its death. This general picture appears to be characteristic of many higher plants. The pea and maize behave similarly, although in maize the adult plant always seems to contain some γ- as well as the predominant δ-tocopherol.

Apart from any implications that these results may have in connection with our understanding of the role of tocopherols in plants, referred to later, they are of interest because of their biosynthetic implications. These studies showed first that it was likely, on kinetic grounds (although not proven) that non-α-tocopherols could be converted to α-tocopherol and probably vice versa, and these changes were particularly prevalent during the processes of maturation, seed germination and seed maturation, respectively. Secondly, in maize seedlings, it was shown that total net synthesis of tocopherols can occur by a dark reaction. Zacharowa,[21] in 1954, had, in fact, already reported a dark synthesis of tocopherol in pea and barley seedlings. It is not clear whether phytol itself or a precursor is involved, especially since Fischer and Bohn[22] found no phytol in etiolated seedlings and concluded that phytol is only synthesized in a light reaction in the presence of chlorophyll. The discovery and formulation of the tocotrienol series suggests that phytol itself is probably not involved, but that a $C_{(20)}$ tetra-unsaturated isoprenoid alcohol might be an intermediate in tocopherol biosynthesis. On this hypothesis, the appearance of non-α-tocopherols in seed might be associated with a deficiency in processes leading to (a) the side-chain saturation and (b) possibly also the methylation of unmethylated tocotrienols. Whether the synthesis of α-tocopherol in the green growing plant is by such a route is another matter. Dam, Glavind and Gabrielson[23] concluded that vitamin K synthesis was probably only incidentally related to the presence of chlorophyll in leaves. Baszynski[24] has recently confirmed many of these findings and, using an *in vitro* system, obtained increased synthesis of α-tocopherol in pea seedlings under the influence of methionine and adenosine triphosphate, thus indicating that nuclear methylation of non-α-tocopherols is a probable stage in the synthesis of α-tocopherol in these seedlings. It should be noted that the results of Kubin and Fink,[19] who found quite large amounts of tocopherols in mushrooms, prove that chlorophyll is not necessary for their synthesis. α-Tocopherol has also been found in some yeasts.[25]

Recently, Booth and Hobson-Frohock[26] have found that remarkable differences exist in the α-tocopherol content of leaves. They found that while evergreen and slow-growing leaves contained up to 450 ppm, fast-growing leaves, such as cress and lettuce, contained only about 7 ppm. (This seems surprising since lettuce has long been believed to be a particularly rich source of vitamin E.[27]). They found that, in general, tocopherol concentrations were highest in parts of leaves where growth was minimal and lowest where growth was maximal. Dying and fallen leaves contained higher contents than green leaves. Winter concentrations were higher than spring concentrations and it was found that, in *Narcissus*, α-tocopherol contents diminished considerably during development of the blooms. Booth and Hobson-Frohock concluded that α-tocopherol concentrations in leaves varied inversely with the growth rates of parts. Since α-tocopherol is known to be rapidly synthesized by growing plants, they suggested that α-tocopherol is used up when the metabolic rate is high. They showed, further, that tocopherol concentrations in higher plants cannot be related to flowering or seed formation, because in some species maxima are reached long after seeding. These observations deserve study because, in animals, high vitamin E concentrations are usually associated with metabolically active tissues. Booth [28] has recently discovered a tocopherol-destroying enzyme in plants that resembles in some respects the carotene-destroying enzyme. The enzyme catalyses the oxidation of tocopherol in chloroplast preparations from damaged leaves and may play a part, perhaps in maintaining tocopherol in the reduced form, in intact leaves. Before leaving this topic, the recent observations of Sironval and Tannir-Lomba[29] must be mentioned. They found that the vitamin E content of *Fragaria vesca* leaves increased markedly with increased day-length and associated this with flower development. This finding contrasts with the general conclusions of Booth and Hobson-Frohock and indeed they differ in other important respects. Thus, Sironval and El Tannir-Lomba found much more "vitamin E" in young leaves than old ones and much more near the apex of leaves than near the base. The quantitative data of the two groups are, in fact, in such contradiction that it is imperative to search for a likely cause, especially since questions of physiological function may depend on it. It seems possible (as was suggested by Booth and Hobson-Frohock) that since Sironval and El Tannir-Lomba did not use any form of differential chromatography in their analyses, they measured not only tocopherols, but a variety of other reducing substances, the concentrations of which might be as variable as is that of α-tocopherol.

FUNCTIONS OF TOCOPHEROLS IN PLANTS

Very little is known about the functions of tocopherols in plants. They are not involved in any known enzyme system as prosthetic groups and Booth's description of a tocopherol oxidizing system is, I believe, the first such discovery in this field. It may be significant that Booth finds both α-tocopherol and the enzyme to be located inside the chloroplasts, whereas he finds γ-tocopherol (when it is present) mainly outside.[28] Booth suggests that α-tocopherol might, therefore, play a part in photosynthesis, perhaps protecting chlorophyll from oxidation. This is certainly one possibility, but what then is the role of tocopherols in non-photosynthesizing plants? It is only recently that methods of analysis have been such that detailed analyses of tocopherols in plant materials can be made. In our laboratories we have found α-tocopherol in mushrooms and yeast, and the recent work of Kubin and Fink[19] shows that certain fungi contain, on a dry weight basis, as much tocopherol as Booth and Hobson-Frohock found in fresh lettuce and many other leaves. The main difference would appear to be the universal finding of ζ-tocopherol in mushrooms, usually as the *predominant* tocopherol. We would like to know whether the substance reported by these workers is a tocol or a tocotrienol. Several workers have indeed suggested that the storage of tocopherols, especially the non-α-tocopherols, in seeds may be related to the necessity of preventing the oxidation of unsaturated fats in these almost *in vitro* systems. Examination of the literature does in fact show a rough correlation between the tocopherol content of seed oils and the degree of unsaturation of the oil. Furthermore, γ- and δ-tocopherol which are *in vitro* the most potent anti-oxidants in the tocopherol series, are mainly found in highly unsaturated seed oils. This may, of course, be fortuitous and, clearly, the seed is quite a different metabolic system to the growing leaf.

REFERENCES

1. H. M. EVANS and K. S. BISHOP, *Science* **14,** 650 (1922).
2. J. GREEN, P. MAMALIS, S. MARCINKIEWICZ and D. MCHALE, *Chemistry and Industry* 73 (1960).
3. S. MARCINKIEWICZ and J. GREEN, *Analyst* **84,** 304 (1959).
4. J. GREEN, unpublished results.
5. D. MCHALE, J. GREEN and S. MARCINKIEWICZ, unpublished results.
6. W. H. HARRISON, J. E. GANDER, E. R. BLAKLY and P. D. Boyer, *Biochim. Biophys. Acta* **21,** 150 (1956).
7. C. MARTINS and H. EILINGSFELD, *Biochem. Z.* **328,** 507 (1957).
8. J. B. MACKENZIE and C. G. MACKENZIE, *J. Nutrition* **72,** 322 (1960).
9. D. MCHALE, private communication.
10. P. W. R. EGGITT and F. W. NORRIS, *J. Sci Food Agr.* **7,** 493 (1956).
11. C. H. LEA, *J. Sci. Food Agr.* **10,** 537 (1959).
12. F. BRO-RASMUSSEN and W. HJARDE, *Acta Chem. Scand.* **11,** 34 (1957).
13. P. W. R. EGGITT and F. W. NORRIS, *J. Sci. Food Agr.* **6,** 689 (1955).
14. J. GREEN, S. MARCINKIEWICZ and P. R. WATT, *J. Sci. Food Agr.* **6,** 274 (1955).
15. M. L. QUAIFE, *J. Biol. Chem.* **175,** 605 (1948).
16. A. EMMERIE and C. ENGEL, *Rec. Trav. Chim.* **57,** 1351 (1938).
17. F. BROWN, *J. Sci. Food. Agr.* **4,** 161 (1953).
18. R. J. WARD, *Brit. J. Nutrition* **12,** 231 (1958).
19. H. KUBIN and H. FINK, *Fette Seifen Anstrichmittel* **63,** 280 (1961).
20. J. GREEN, *J. Sci. Food Agr.* **9,** 801 (1958).
21. M. P. ZACHAROWA, *Trudy WNIWI* **5,** 159 (1954).
22. F. G. FISCHER and H. BOHN, *Ann.* **611,** 224 (1958).
23. H. DAM, J. GLAVIND and E. K. GABRIELSON, *Acta Physiol. Scand.* **13,** 9 (1947).
24. T. BASZYNSKI, *Acta Soc. Botanicorum Piloniae* **30,** 307 (1961).
25. A. T. DIPLOCK, J. GREEN, E. E. EDWIN and J. BUNYAN, *Nature, Lond.* **189,** 749 (1961).
26. V. H. BOOTH and A. HOBSON-FROHOCK, *J. Sci. Food Agr.* **12,** 253 (1960).
27. H. R. ROSENBERG, *Chemistry and Physiology of the Vitamins*, Interscience, New York (1942).
28. V. H. BOOTH, personal communication.
29. C. SIRONVAL and J. EL TANNIR-LOMBA, *Nature, Lond.* **185,** 855 (1960).

CHAPTER 6

THE ROLE OF UBIQUINONE AND PLASTOQUINONE IN PLANT METABOLISM

F. W. HEMMING and R. A. MORTON

Department of Biochemistry, The University, Liverpool

THIS paper was prepared with three aims in mind. In the first place an attempt will be made to review briefly the present state of knowledge concerning the probable functions of plastoquinone and ubiquinone in living cells. This will involve a discussion of the biochemical significance of their isomeric, cyclized phenols. (For more detailed reviews see references 1–7.) Secondly, possible connections of the biosynthetic pathways of these quinones with those of other polyisoprenoid compounds will be discussed. Some of the points raised will then be illustrated by referring to recent work on the constituents of the spadices of *Arum maculatum* and of the green leaves of the tobacco plant.

PLASTOQUINONE, SOLANACHROMENE AND SOLANESOL

Just sixteen years ago Kofler,[8] working in Switzerland isolated a benzoquinone from a petroleum ether extract of dried lucerne. The problem of

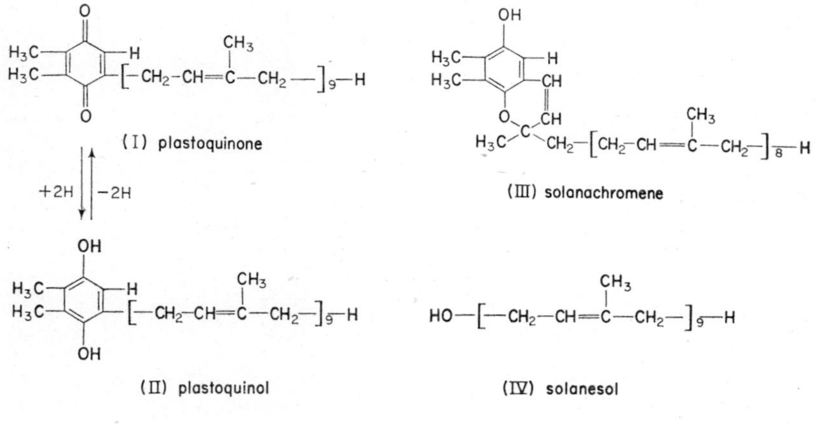

FIG. 1.

the precise structure of this compound was not taken up again until 1958, when workers in Switzerland[9–12] and in the United States[13,14] presented conclusive evidence for structure I (Fig. 1). Prior to this chemical work the compound had been rediscovered by Crane. He called it plastoquinone because distribution studies showed that it occurred mainly in chloroplasts.[15] The name koflerquinone is also used.

The long isoprenoid side-chain confers lipid solubility on the molecule. As a *p*-benzoquinone it can be reduced to the quinol (structure II, Fig. 1). This reaction is readily reversible and is important for two reasons. In the first place it is upon this property that theories concerning its role in

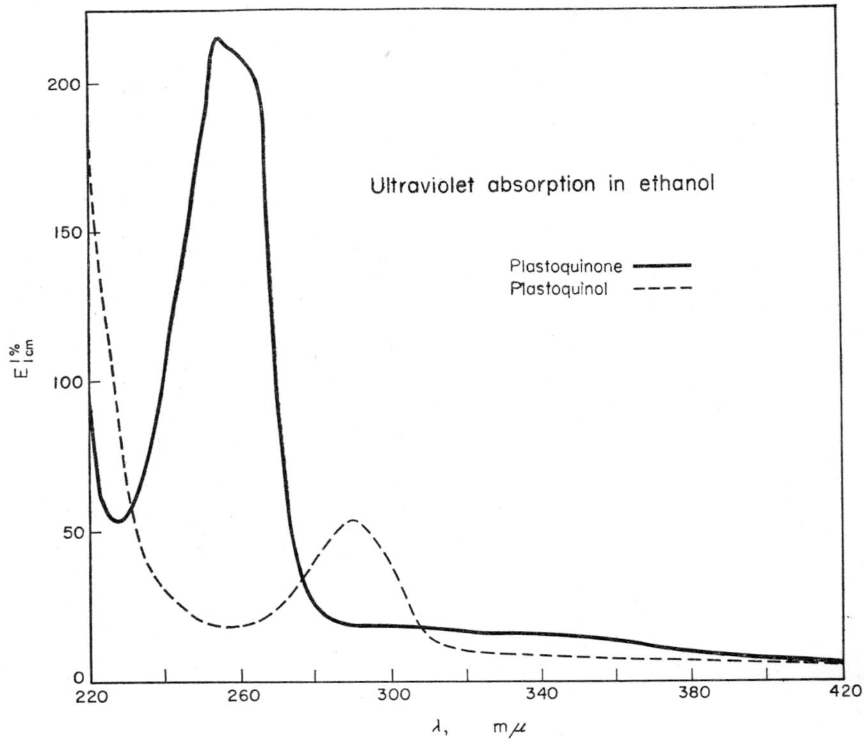

FIG. 2. Ultraviolet absorption of plastoquinone and plastoquinol.

chloroplasts are based. In the second place it provides a convenient method of assay. The quinone shows maximal ultraviolet absorption at 255 mμ (in ethanol) and only slight absorption at 290 mμ, whereas the quinol exhibits only slight absorption at 255 mμ and maximal absorption at 290 mμ (Fig. 2). A reducing agent will, therefore, produce changes in ultraviolet absorption which can be measured accurately with the help of a spectrophotometer. If one wishes to assay the quinol a weak oxidizing agent (e.g. ferric chloride in ethanol) is used and the reverse changes are followed. This is the main method by which the levels of the quinone and quinol in tissue extracts are measured. The 2′-3′ double bond in the isoprenoid side chain permits the molecule to cyclize to the structure III (Fig. 1). The compound so formed is a phenol with a chromene ring and is an isomer of plastoquinone. It has in fact been isolated from aged, flue-cured tobacco by Rowland.[16] The important question as to whether the cyclization occurs in the living tobacco plant or during the processing of the leaf is still not answered. Rowland called this compound solanachromene because two years earlier he had isolated from flue-cured tobacco, the related isoprenoid alcohol, solanesol (structure IV, Fig. 1).[11,13,17] Leaves of many other plants yield little, if any, of these two compounds, and it is surprising that no source as good as tobacco has yet been found.

UBIQUINONE AND UBICHROMENOL

The early work on ubiquinone was done in Liverpool. In 1950, Festenstein[18] isolated the compound from the mucosal lining of horse intestine.

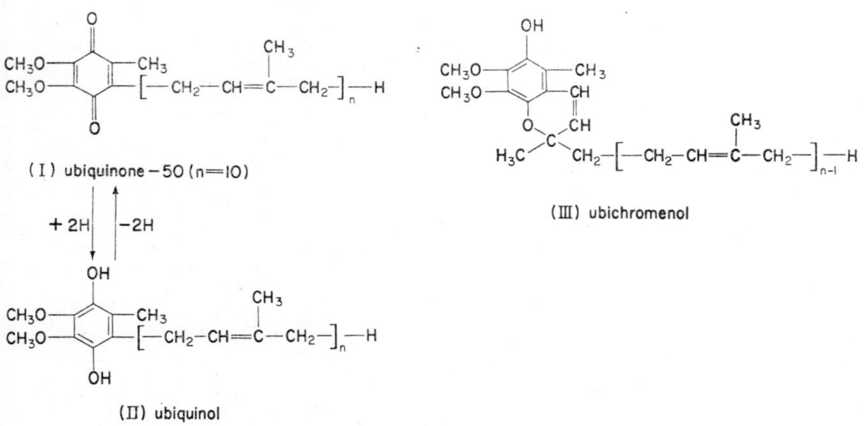

FIG. 3.

This was followed by studies on its distribution and of changes in animal tissue levels following changes in nutritional status of the animals. Then the chemistry of the compound isolated from pig heart was considered and structure I (Fig. 3) was proposed early in 1958.[19] Collaboration with Swiss workers led by Isler confirmed this later in the year.[20,21] At the same time the joint efforts of Green's group in Madison and Folkers' group in Rahway, New York, led to the same answer.[22–24] The European groups called their compound ubiquinone because this type of quinone appeared to be ubiquitous. The American groups settled for coenzyme Q because they believed it to be a catalyst in electron transport. Like plastoquinone it is a lipid-soluble benzoquinone, but differs in that there are methoxyl groups in positions 2 and 3 and a methyl group in position 6. It has characteristic ultraviolet absorption (λ_{max} 275 mμ in ethanol) and can be assayed by a method similar to that for plastoquinone (see Fig. 4). The corresponding quinol (structure II, Fig. 3) has λ_{max} 290 mμ in ethanol. The cyclic isomer (structure III, Fig. 3) has been found in most animal tissues and in some microorganisms.[3,25–32] The structure of this phenol

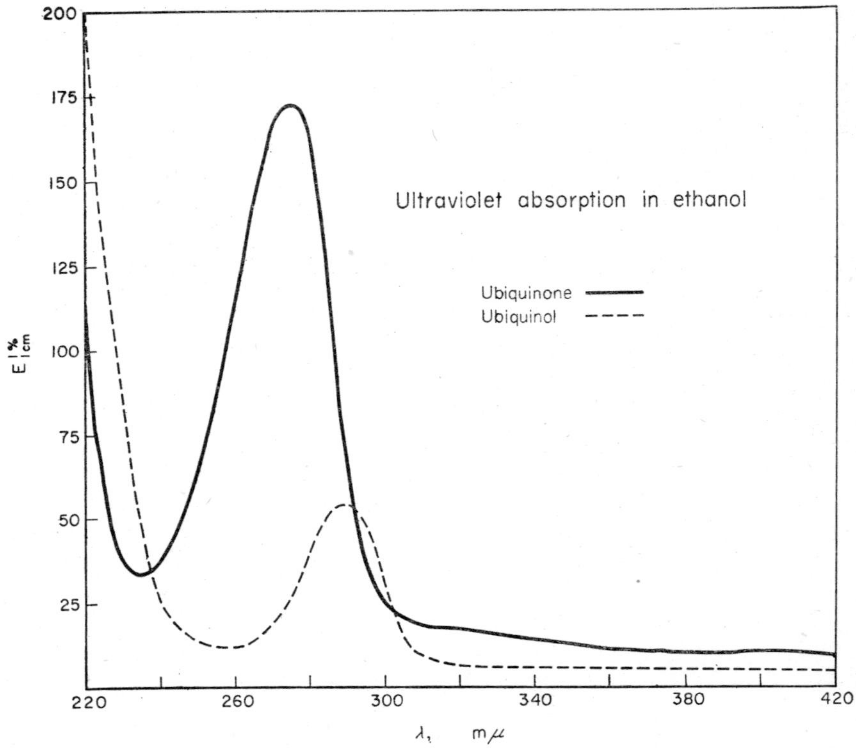

FIG. 4. Ultraviolet absorption of ubiquinone-50 and ubiquinol-50.

was established by Morton's group in 1959.[33,34] It was given the name ubichromenol.

Ubiquinone has been found in all animal tissues that have been investigated. It is also present in plants and in most microorganisms. Also, at this stage in the work it became necessary to consider ubiquinone as a generic term for it was clear that a series of compounds existed, differing only in the number of isoprenoid residues in the side chain. Five homologues (or more exactly isoprenologues) are known to occur naturally. In Fig. 3, *n* can be any number from 6 to 10. In most animals and plants *n* is equal to 10 but occasionally *n* equals 9 (e.g. rat and mouse tissues, wheat germ oil and corn oil). The lower isoprenologues (where $n = 6, 7$ or 8) are probably formed only in microorganisms.[2,4,35–38]

As a rough generalization one can say that the ubiquinone level reflects the respiratory rate of the tissue concerned. There are, however, important exceptions to this in some of the bacteria.

UBIQUINONE IN MITOCHONDRIAL ELECTRON TRANSPORT

The ubiquinone of animal and plant tissues is to be found concentrated in the mitochondria.[15,39–45] Most of the work on its role has been carried

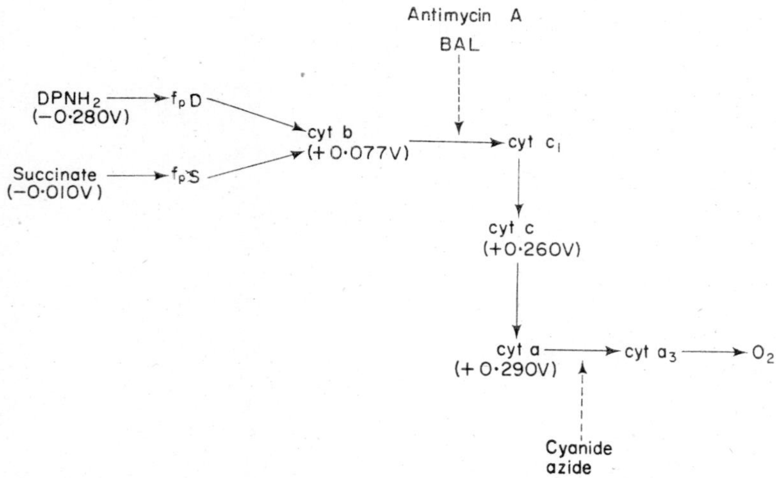

FIG. 5. The phosphorylating electron transport system including redox potentials and sites of action of inhibitors (Redfearn[46]).

out on mitochondria of animal origin. However, it seems reasonable to assume that functionally the mitochondria from plant tissues are basically similar to those from animals.

Early investigations were led by Green's group in Wisconsin. They worked with ox-heart mitochondria and a number of smaller particles derived from these. Most of the preparations were non-phosphorylating but they were capable of passing electrons from an oxidizable substrate along the chain of cytochromes. Thus succinate could be oxidized to fumarate by removing electrons and passing them along the chain of cytochromes to molecular oxygen (Fig. 5). The liberated hydrogen ions combined with the two electrons and a half molecule of oxygen to form a molecule of water. Other compounds (derived principally from carbohydrates) could be oxidized by passing their hydrogen atoms on to diphosphopyridine nucleotide (DPN, NAD) which was then oxidized in the same way as succinate. This transport of electrons could be blocked by inhibitors acting at points indicated by the broken-line arrows. During this passage of electrons both endogenous and exogenous ubiquinone was first reduced to the quinol and then oxidized back to the quinone.[3-6] It then became necessary to investigate the possibility that ubiquinone is a component of the electron transport chain. The problem is still not completely resolved, but the evidence accumulated by many workers has led to a number of reviews,[2-6,45,46] most of which make the following points.

Firstly, there is sufficient ubiquinone in the mitochondria for it to function as an electron carrier. On a molar basis there is twelve times more ubiquinone than cytochrome c_1 in pig heart mitochondria[46,47] and ten times more in rat liver mitochondria.[46,48] In *Arum* spadix mitochondria, the corresponding ratio is 6:1.[47] To put these figures in perspective one should remember that in rat liver mitochondria the ratio for flavin to cytochrome is 3·6:1 and for pyridine nucleotide to cytochrome is 19:1.[5] The molar concentration of ubiquinone is thus intermediate between that of flavin and pyridine nucleotide.

Secondly, if ubiquinone is extracted from mitochondria with a suitable organic solvent the treated mitochondria no longer oxidize substrates such as succinic acid. The activity can be restored in a specific manner by adding ubiquinone at a concentration comparable with that found in the untreated mitochondria.[3,49,50]

Thirdly, the redox potential of ubiquinone at the pH of the cell has been calculated to be +0·120 V.[51] This is intermediate between that of cytochrome c (+0·260 V) and cytochrome b (+0·077 V).[52]

Fourthly, both succinate and DPNH (NADH) will act as substrates for ubiquinone reductase and the reduction involved is insensitive to antimycin A and cyanide. If ubiquinone had been to the right of cytochrome c_1 in Fig. 5, then antimycin A would have blocked the reduction. Oxidation of ubiquinol by oxygen is, however, inhibited by both of these substances.[4,53,54] Since on the basis of other investigations antimycin A is known to inhibit the electron transport chain just in front of cytochrome c_1 and in view of the redox potential of ubiquinone, it seems that ubiquinol probably hands on its electrons to cytochrome c_1, i.e. it is positioned to the left of cytochrome c_1 in Fig. 5. As one would expect on the basis of this location, exogeneous cytochrome c_1 is reduced when ubiquinol is added to cyanide-poisoned preparations.[3]

In general these points are accepted by most workers in the field. More crucial, however, are the rates of reduction of endogenous ubiquinone and of oxidation of endogenous ubiquinol in the mitochondria. If the rates are equal to or higher than the overall rate of electron transport along the chain then the evidence for putting ubiquinone in the pathway becomes very strong. If, in fact, the rates are lower, consideration of alternative hypotheses becomes necessary. It is on this point that sharpest differences of opinion are to be found. Green's evidence[3,4] indicates that the enzymic rates are of the correct order of magnitude while other workers, including Chance and Redfearn,[55,56] have provided evidence that they are far too low. For this reason Green's scheme ((i) Fig. 6) includes ubiquinone on the main chain while other schemes have ubiquinone in a blind alley ((ii) Fig. 6) or on an alternative branch pathway ((iii) Fig. 6).

UBIQUINONE IN OXIDATIVE PHOSPHORYLATION

On the whole the above schemes have been worked out using non-phosphorylating preparations. An important aspect of the passage of

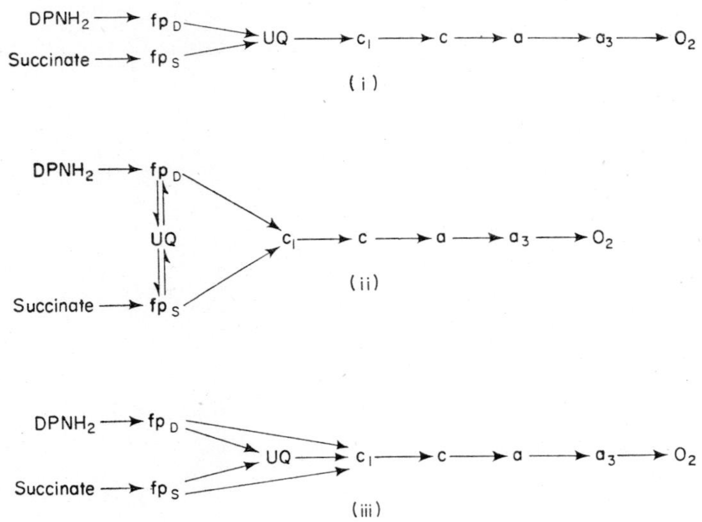

FIG. 6. Three possible schemes for the participation of ubiquinone (UQ) in the non-phosphorylating electron-transport system (Redfearn[45]).

electrons down the energy gradient of the electron transport chain is the trapping in the form of adenosine triphosphate of the energy so released. This is the essence of oxidative phosphorylation, and it involves the condensation of inorganic phosphate ultimately with adenosine diphosphate. Since the study of ubiquinone in electron transport *per se* is at present rather in stalemate, it is natural to turn attention to the problems of oxidative phosphorylation. Ubiquinone is again being considered. In the first place there is evidence that in phosphorylating systems inorganic phosphate will reduce the rate of oxidation of ubiquinol whereas ADP reverses the effect.[57] (See also Redfearn and Pumphrey.[46,48]) Secondly, the idea that a derivative of ubiquinone could be phosphorylated, and then by various molecular rearrangements become a high energy phosphate, has proved attractive to many workers.[58–67] This derivative would then hand on its phosphate to adenosine diphosphate. However, the hypothesis is based more upon organic chemistry than enzymology. Intermediates that have been considered are ubiquinol and the phenols ubichromenol and ubichromanol.

Thirdly, the work of Brodie with *Mycobacterium Phlei* indicates clearly that, in this organism, the naphthoquinone, vitamin K_1, functions in oxidative phosphorylation. Actively phosphorylating preparations yielded an intermediate which was later shown to be a naphthochromanol. The

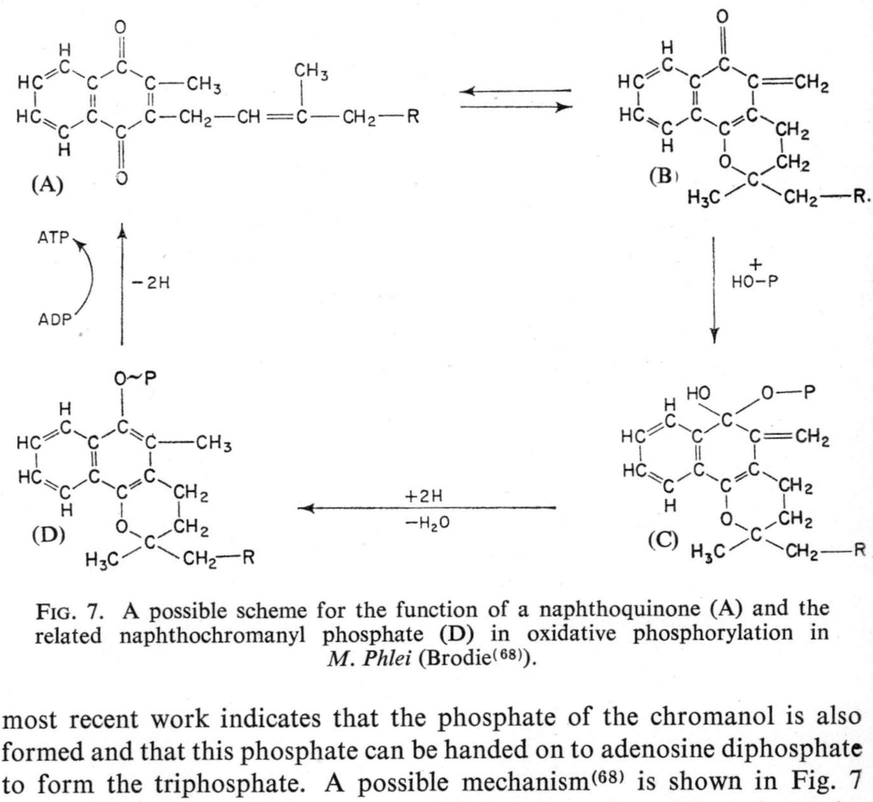

FIG. 7. A possible scheme for the function of a naphthoquinone (A) and the related naphthochromanyl phosphate (D) in oxidative phosphorylation in *M. Phlei* (Brodie[68]).

most recent work indicates that the phosphate of the chromanol is also formed and that this phosphate can be handed on to adenosine diphosphate to form the triphosphate. A possible mechanism[68] is shown in Fig. 7 and is based on that of Chmielewska.[63] Structures A and B are isomeric. C is formed as a result of charge splitting at carbon atom 1 and attack by inorganic phosphate. This is rapidly reduced and water removed to yield the high energy phenyl phosphate D (naphthochromanyl phosphate). Oxidation of D to A is accompanied by the phosphorylation of adenosine diphosphate. The net effect is the coupling of inorganic phosphate with adenosine diphosphate to give the high energy adenosine triphosphate. This involves the formation of a high energy phenyl phosphate as intermediate, the energy being derived by reduction of C to D and passed on to ATP during the oxidation of D to the original quinone. The cycle can then be repeated. As yet this is one of a number of hypotheses and there is little experimental evidence to justify making a choice between them, but it does indicate a possible role of phenyl phosphates in oxidative phosphorylation.

Although ubiquinones in mitochondria from plants or animals might be analogous to vitamin K in *M. phlei*, there is no direct evidence to implicate their phenol derivatives (ubichromanol and ubichromenol) in oxidative phosphorylation. However, it may be relevant to remember that other phenols such as thyroxine and 2,4-dinitrophenol uncouple phosphorylation from electron transport. Administration of these, also causes changes in animal tissue levels of ubiquinone and ubichromenol.[69]

PLASTOQUINONE IN PHOTOSYNTHETIC ELECTRON TRANSPORT

It is now almost generally accepted that in animals vitamin K_2 as such is unlikely to be involved in electron transport if only because there seems to be very little in the tissues. The same can probably be said for the chloroplasts, which contain little vitamin K_1 but yield a relatively large amount of plastoquinone.[70] On a molar basis there is twenty to forty times more plastoquinone than cytochrome f in sugar-beet chloroplasts.[71]

The function of plastoquinone has received less attention than that of ubiquinone. That it is associated mainly with the chloroplasts was first shown by Crane in 1959.[15] Differential centrifugation of a homogenate of spinach leaves yielded a number of fractions. It was found that the distribution of plastoquinone in these followed closely that of chlorophyll, whereas that of ubiquinone followed that of succinic dehydrogenase. Also the green and white stripes of the leaves of *Pandanus vetchii*, and the white roots and green shoots of seven-day old corn seedlings were analysed. It was shown that plastoquinone was concentrated mainly in the green tissues whereas ubiquinone was more evenly distributed.

A number of schemes to explain the known facts of photosynthesis are discussed in the current literature. The scheme in Fig. 8 is based on that proposed by Arnon.[72] The main object is to maintain the photolysis of water. Two electrons from the hydroxyl group of water are accepted by the pigment which has been activated by two photons. Activation involves removal of two electrons ($2e^{-}a$) from the pigment to acceptor A. They are passed on to cytochrome f and then to chlorophyll, and at this stage a molecule ATP is generated. For chlorophyll to accept the electron they have previously to be activated by two photons, the liberated electron ($2e^{-}a$) being passed on to acceptor B^{+}. Normally, B^{+} is triphosphopyridine nucleotide, (TPN, NADP), and the net effect is that one molecule of ATP is generated for each molecule of TPNH and each half-molecule of oxygen generated.

This train of events is called "non-cyclic" photophosphorylation. The TPNH and ATP are used mainly for the conversion of carbon dioxide and water to starch and so the energy of the photons is captured and stored. Not all of the TPNH is utilized for assimilation of CO_2 and some passes its electrons on to the cytochrome f via the flavines. This is accompanied by a phosphorylation and so a cyclic photophosphorylation also occurs.

In vitro one can replace TPN as electron acceptor B^+ by a number of synthetic dyes and then study the photo-induced transport of electrons more easily. In this way it has been possible to show that light-sustained electron transport ceases in isolated sugar-beet chloroplasts after removal

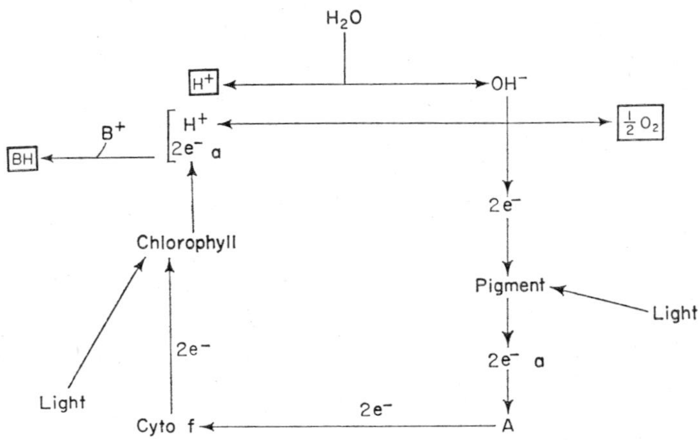

FIG. 8. A possible scheme for light-induced electron transport in chloroplasts (Arnon[72]).

of plastoquinone and is restored by replacement of the quinone.[70,73] Also when isolated chloroplasts from sugar-beet or spinach were illuminated, the ratio of plastoquinol to plastoquinone increased.[71,74] It seems probable on the basis of this, that plastoquinone is an electron acceptor in the photosynthetic electron transport chain. Recent work suggests that it functions as acceptor A.[90]

Nothing is known concerning the possible role of plastoquinone in linking this electron transport to phosphorylation. Hypotheses similar to those for oxidative phosphorylation have been considered. It could be that solanachromene (the cyclized phenolic isomer of plastoquinone) has a function here.

BIOSYNTHESIS

Both quinones have isoprenoid side chains and in animals there is little doubt that the side chain of ubiquinone can be derived from mevalonate.[75,76] The work of Olson[77–79] shows that in the rat the quinone moiety is at least partly derived from the aromatic ring of phenylalanine. The experiments of Martius[80,81] indicate that the final step occurs in the mitochondria and is the condensation of solanesyl pyrophosphate with 2,3 dimethoxy, 6-methyl-1,4-benzoquinone as in Fig. 9, pyrophosphate being eliminated.

In most animal tissues the natural quinone is that with ten isoprene residues in its side chain, i.e. ubiquinone-50. For some time now we have been looking in animal tissues for solanesol-50, the expected precursor of the side chain of ubiquinone-50. So far we have had no success. We did,

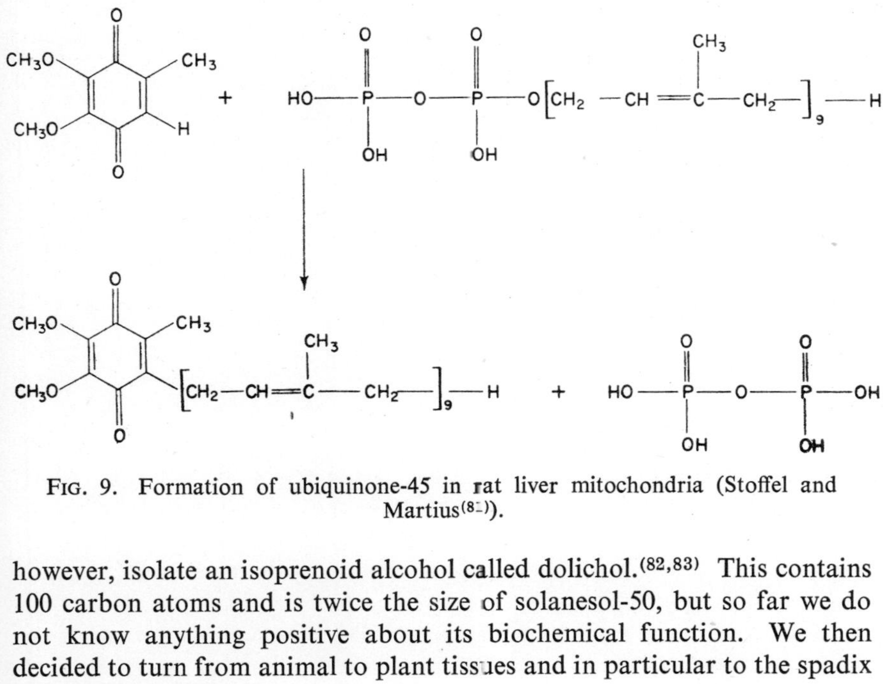

FIG. 9. Formation of ubiquinone-45 in rat liver mitochondria (Stoffel and Martius[81]).

however, isolate an isoprenoid alcohol called dolichol.[82,83] This contains 100 carbon atoms and is twice the size of solanesol-50, but so far we do not know anything positive about its biochemical function. We then decided to turn from animal to plant tissues and in particular to the spadix of *Arum maculatum*.

THE SPADIX OF *ARUM MACULATUM*

This tissue is capable of an extremely high respiratory rate[84] and contains a fairly high concentration of cytochromes.[85,86] It contains little or no chlorophyll and presumably shows little photosynthetic activity.

Spadices (2 kg) were collected mainly at Scarborough but also at Loggerheads, near Mold in North Wales.[87] The unsaponifiable lipid was chromatographed on alumina and the fractions subjected to more detailed chromatographic procedures to yield 0·7 μg plastoquinone-45 and 32·2 μg ubiquinone-50 per g tissue (Table 1). This fitted in with the low photosynthetic and high respiratory activity of the tissue. We also found 20 μg/g of a mixture of isoprenoid alcohols. This did not contain detectable amounts of solanesol-45 but did contain 5 μg/g solanesol-50. This correlation indicated that the solanesol-50 might well be acting as the precursor of the side chain of ubiquinone-50 in this tissue. It was the first time that solanesol-50 has been isolated from natural sources. When we extended our work to fresh green tobacco leaves the reverse situation held.(88) The tissue (1 kg) yielded little ubiquinone-50 and no detectable solanesol-50 or dolichol-100 but did provide plastoquinone-45 (330 μg/g) and solanesol-45 (55 μg/g). The two compounds were concentrated in the chloroplasts

TABLE 1. SOME LIPID CONSTITUENTS OF PLANT TISSUES
(μg/g wet wt.)

Constituent	Green leaves of tobacco	Spadices of *Arum maculatum*
Plastoquinone	356	0·7
Solanesol	64	N.D.
Ubiquinone-50	trace	32·2
Solanesol-50	N.D.	5

N.D. = not detected.

(Table 2). It seems, then, that in plant tissues solanesols (or their pyrophosphates) might well be the precursors of the side chains of these two quinones as in Martius' scheme for ubiquinone in animal tissues, and that in fresh tobacco leaves the final stage in the biosynthesis of plastoquinone occurs in the chloroplasts.

Finally, in considering the biochemistry of these compounds in green leaves one must take into account the phenomenon of accumulation. Booth and Hobson-Frohock have drawn attention to this phenomenon in the case of tocopherols.(89) The concentrations of tocopherols in various leaves were inversely proportional to the growth rate and rose rapidly as the leaf aged. In deciduous leaves, levels were very much higher in autumn than in spring. Kofler has drawn attention to this in the case of plastoquinone.(10) We have been able to confirm this and have also met the phenomenon in the case of the isoprenoid alcohols. In leaves collected in autumn, levels of these compounds were again much higher than in early summer or in spring. At first glance, it looks to be a wasteful process for presumably in late autumn the photosynthetic activity of deciduous leaves is low. On the other hand it does make the isolation and chemical identification of compounds such as I have been describing that much easier.

CONCLUSION

However valid are the various theories relating to the finer points of the function of ubiquinone and plastoquinone, there is no doubt that these two relatively simple compounds play a central role in the metabolism of plants and animals. Without plastoquinone green plants would be unable to trap the energy of sunlight and without ubiquinone the utilization of this trapped energy in both animal and plant kingdoms would be inefficient, to say the least.

TABLE 2. THE INTRACELLULAR DISTRIBUTION OF PLASTOQUINONE, SOLANESOL AND PROTEIN IN FRESH GREEN TOBACCO LEAVES

Fraction	Plastoquinone (μ moles/kg/wet wt.)	Solanesol (μ moles/kg/wet wt.)	Protein (g/kg wet wt.)
Chloroplasts	442	86	29·04
Mitochondria	0	0·4	1·96
Supernatant	34	15	26·44

REFERENCES

1. R. A. MORTON, *Nature, Lond.* **182,** 1764 (1958).
2. R. A. MORTON, *Vitamins and Hormones* **19,** 1 (1961).
3. *Quinones in Electron Transport,* Ciba Foundation Symposium, eds. G. E. W. Wolstenholme and C. M. O'Connor (1961).
4. T. RAMASARMA, *J. Sci. Ind. Res.* **20A,** 492 (1961).
5. D. E. GREEN, *Adv. Enzymol.* **21,** 73 (1959).
6. D. E. GREEN and R. L. LESTER, *Fed. Proc.* **18,** 987 (1959).
7. P. A. VAN ZWIETEN, *Chem. Weekblad* **56,** 257 (1960).
8. M. KOFLER, *Festschrift für Emil Christoph Barell,* Basel p. 199 (1946).
9. C. v. PLANTA, E. BILLETER and M. KOFLER, *Helv. Chim. Acta* **42,** 1278 (1959).
10. M. KOFLER, A. LANGEMANN, R. RÜEGG, L. H. CHOPARD-DIT-JEAN, A. RAYROUD and O. ISLER, *Helv. Chim. Acta* **42,** 1283 (1959).
11. M. KOFLER, A. LANGEMANN, R. RÜEGG, U. GLOOR, U. SCHWIETER, J. WÜRSCH, O. WISS and O. ISLER, *Helv. Chim. Acta* **42,** 2252 (1959).
12. M. KOFLER, A. LANGEMANN and R. RÜEGG, *Chimia (Switz.)* **13,** 115 (1959).
13. R. E. ERICKSON, C. H. SHUNK, N. R. TRENNER, B. H. ARISON and K. FOLKERS, *J. Amer. Chem. Soc.* **81,** 4999 (1958).
14. C. H. SHUNK, R. E. ERICKSON, E. L. WONG and K. FOLKERS, *J. Amer. Chem. Soc.* **81,** 5000 (1959).
15. F. L. CRANE, *Plant Physiol.* **34,** 128 (1959).
16. R. L. ROWLAND, *J. Amer. Chem. Soc.* **80,** 6130 (1958).
17. R. L. ROWLAND, P. H. LATIMER and J. A. GILES, *J. Amer. Chem. Soc.* **78,** 4680 (1956).
18. G. N. FESTENSTEIN, Ph.D. Thesis, University of Liverpool (1950).
19. N. I. FAHMY, F. W. HEMMING, R. A. MORTON, J. Y. F. PATTERSON and J. F. PENNOCK, *Biochem. J.* **70,** 1P (1958).

20. R. A. MORTON, U. GLOOR, O. SCHINDLER, G. M. WILSON, L. H. CHOPARD-DIT-JEAN, F. W. HEMMING, O. ISLER, W. M. F. LEAT, J. F. PENNOCK, R. RÜEGG, U. SCHWIETER and O. WISS, *Helv. Chim. Acta* **41,** 2343 (1958).
21. U. GLOOR, O. ISLER, R. A. MORTON, R. RÜEGG and O. WISS, *Helv. Chim. Acta* **41,** 2357 (1958).
22. D. E. WOLF, C. H. HOFFMAN, N. R. TRENNER, B. H. ARISON, C. H. SHUNK, B. O. LINN, J. F. MCPHERSON and K. FOLKERS, *J. Amer. Chem. Soc.* **80,** 4752 (1958).
23. C. H. SHUNK, B. O. LINN, E. L. WONG, P. E. WITTREICH, F. M. ROBINSON and K. FOLKERS, *J. Amer. Chem. Soc.* **80,** 4753 (1958).
24. R. L. LESTER, F. L. CRANE and Y. HATEFI, *J. Amer. Chem. Soc.* **80,** 4751 (1958).
25. A. T. DIPLOCK, J. GREEN, E. E. EDWIN and J. BUNYAN, *Nature, Lond.* **189,** 749 (1961).
26. E. E. EDWIN, A. T. DIPLOCK, J. BUNYAN and J. GREEN, *Biochem. J.* **79,** 91 (1961).
27. J. GREEN, A. T. DIPLOCK, J. BUNYAN and E. E. EDWIN, *Biochem. J.* **79,** 108 (1961).
28. A. T. DIPLOCK, J. BUNYAN, J. GREEN and E. E. EDWIN, *Biochem. J.* **79,** 105 (1961).
29. N. F. CUNNINGHAM and R. A. MORTON, *Biochem. J.* **72,** 92 (1958).
30. J. S. LOWE, R. A. MORTON and J. VERNON, *Biochem. J.* **67,** 228 (1957).
31. L. MERVYN and R. A. MORTON, *Biochem. J.* **72,** 106 (1959).
32. N. M. PACKTER and J. GLOVER, *Nature, Lond.* **187,** 413 (1960).
33. D. L. LAIDMAN, R. A. MORTON, J. Y. F. PATERSON and J. F. PENNOCK, *Chem. and Ind.* p. 1019 (1959).
34. D. L. LAIDMAN, R. A. MORTON, J. Y. F. PATERSON and J. F. PENNOCK, *Biochem. J.* **74,** 541 (1960).
35. R. L. LESTER and F. L. CRANE, *J. Biol. Chem.* **234,** 2169 (1959).
36. A. C. PAGE, B. O. LINN, P. H. GALE, E. L. WONG, C. H. SHUNK and K. FOLKERS, *Fed. Proc.* **18,** 1176 (1959).
37. K. P. PANOYA, D. H. L. BISHOP and H. K. KING, *Biochem. J.* **78,** 35P (1961).
38. A. C. PAGE, P. H. GALE, F. R. KONIUSZY and K. FOLKERS, *Arch. Biochem. Biophys.* **85,** 474 (1959).
39. F. W. HEMMING, J. F. PENNOCK and R. A. MORTON, *Biochem. J.* **68,** 29P (1958).
40. F. W. HEMMING, J. F. PENNOCK and R. A. MORTON, *Biochem. J.* in the press.
41. E. C. SLATER, H. RUDNEY, J. BOUMAN and J. LINKS, *Biochim. Biophys. Acta* **47,** 497 (1961).
42. A. S. AIYAR and A. SREENIVASAN, *Biochem. J.* **82,** 182 (1962).
43. A. S. AIYAR and A. SREENIVASAN, *Biochem. J.* **82,** 179 (1962).
44. P. S. SASTRY, J. JAYARAMAN and T. RAMASARMA, *Nature, Lond.* **189,** 577 (1961).
45. E. R. REDFEARN, *Ann. Reports Chem. Soc.* **57,** 395 (1960).
46. E. R. REDFEARN, *Biological Structure and Function, Proc. First IUB/IUBS International Symposium*, Vol. 11, p. 181, eds. T. W. Goodwin and O. Lindberg, Academic Press, London (1961).
47. A. M. PUMPHREY and E. R. REDFEARN, *Biochem. J.* **76,** 61 (1960).
48. E. R. REDFEARN and A. M. PUMPHREY, *Biochem. Biophys. Res. Comm.* **3,** 650 (1960).
49. R. L. LESTER and S. FLEISCHER, *Arch. Biochem. Biophys.* **80,** 470 (1959).
50. R. L. LESTER and S. FLEISCHER, *Biochim. Biophys. Acta* **47,** 358 (1961).
51. C. D. JOEL, M. L. KORNOVSKY, E. G. BALL and O. COOPER, *J. Biol. Chem.* **233,** 1565 (1958).
52. J. COLPA-BOONSTRA and F. A. HOLTON, *Biochem. J.* **32,** 4P (1959).
53. F. L. CRANE, Y. HATEFI, R. L. LESTER and C. WIDMER, *Biochim. Biophys. Acta* **25,** 220 (1957).
54. A. M. PUMPHREY, E. R. REDFEARN and R. A. MORTON, *Chem. and Ind.*, p. 178 (1958).
55. E. R. REDFEARN and A. M. PUMPHREY, *Biochem. J.* **76,** 64 (1960).
56. B. CHANCE and E. R. REDFEARN, *Biochem. J.* **80,** 632 (1961).
57. Y. HATEFI, *Biochim. Biophys. Acta* **31,** 502 (1959).
58. J. S. C. WESSELS, *Rec. Trav. Chim.* **73,** 529 (1954).
59. K. HARRISON, *Nature, Lond.* **181,** 1131 (1958).
60. V. M. CLARK, G. W. KIRBY and A. R. TODD, *Nature, Lond.* **181,** 1650 (1958).
61. T. WIELAND and F. PATTERMANN, *Angew. Chem.* **70,** 313 (1958).

62. A. F. BRODIE and J. BALLANTINE, *J. Biol. Chem.* **235**, 232 (1960).
63. I. CHMIELEWSKA, *Biochim. Biophys. Acta* **39**, 170 (1960).
64. E. C. SLATER, *Proc. 4th Int. Congr. Biochem.* Vol. 2, p. 316 (1958).
65. P. D. BOYER, in *The Enzymes*, 2nd Ed., Vol. 3, p. 353, eds. P. D. Boyer, H. Lardy and K. Myrbäck, Academic Press, New York (1960).
66. R. D. DALLAM, *Biochem. Biophys. Res. Comm.* **4**, 106 (1961).
67. H. I. HADLER, *Experientia* **17**, 268 (1961).
68. A. F. BRODIE, *Fed. Proc.* **20**, 995 (1961).
69. E. E. EDWIN, J. GREEN, A. T. DIPLOCK and J. BUNYAN, *Nature, Lond.* **186**, 725 (1960).
70. N. I. BISHOP, *Proc. Nat. Acad. Sci.* **45**, 1696 (1959).
71. E. R. REDFEARN and J. FRIEND, *Nature, Lond.* **191**, 806 (1961).
72. D. I. ARNON, *Fed. Proc.* **20**, 1012 (1961).
73. J. FRIEND and E. R. REDFEARN, *Biochem. J.* **82**, 13P (1961).
74. F. L. CRANE, B. EHRLICH and L. P. KEGEL, *Biochem. Biophys. Res. Comm.* **3**, 37 (1960).
75. U. GLOOR and O. WISS, *Arch. Biochem. Biophys.* **83**, 216 (1959).
76. U. GLOOR and O. WISS, *Biochem. Biophys. Res. Comm.* **2**, 222 (1960).
77. R. E. OLSON and G. H. DIALAMEH, *Biochem. Biophys. Res. Comm.* **2**, 198 (1960).
78. R. E. OLSON and G. H. DIALAMEH, *Fed. Proc.* **19**, 220 (1960).
79. R. E. OLSON, R. BENTLEY, G. H. DIALAMEH and P. GOLD, *Biochem. J.* **82**, 14P (1961).
80. C. MARTIUS, *Amer. J. Clin. Nutr.* **9**, 97 (1961).
81. W. STOFFEL and C. MARTIUS, *Biochem.* **333**, 440 (1960).
82. F. W. HEMMING, J. F. PENNOCK and R. A. MORTON, *Biochem. J.* **74**, 38P (1960).
83. J. F. PENNOCK, F. W. HEMMING and R. A. MORTON, *Nature, Lond.* **186**, 554 (1960).
84. W. O. JAMES and H. BEEVERS, *New Phytol.* **49**, 353 (1950).
85. D. S. BENDALL, *Biochem. J.* **70**, 381 (1958).
86. E. W. SIMON, *J. Exp. Bot.* **8**, 20 (1957).
87. J. F. PENNOCK, F. W. HEMMING and R. A. MORTON, *Biochem J.* **82**, 11P (1961).
88. J. STEVENSON, unpublished work (1961).
89. V. H. BOOTH and A. HOBSON-FROHOCK, *J. Sci. Food Agric.*, p. 251 (1961).
90. E. R. REDFEARN, private communication (1962).

CHAPTER 7

HYDROLYTIC ENZYMES

J. B. PRIDHAM

Chemistry Department, Royal Holloway College, University of London, Englefield Green, Surrey

THE HYDROLYTIC cleavage of covalent bonds such as C—O (esters and glycosides), C—N (proteins and peptides) and O—P (phosphates) is catalysed by a group of hydrolytic enzymes, or hydrolases. In many instances, these enzymes can also effect synthesis by a reversal of their normal hydrolytic action.

This review is intended to cover in general terms, the reaction of plant hydrolases with phenolic compounds and their derivatives. No enzymes in this group have so far been discovered which will react with the aromatic nuclei of phenolics. In fact, the hydrolytic cleavage of C—C bonds is rare and only occurs with some aliphatic carbon chains.[1] Hydrolases will, however, react with substituted phenolic hydroxyl groups and other substituted functional groups which may be present in phenolic compounds. Thus suitable phenolic derivatives will act as substrates for aryl sulphatases, esterases and glycosidases.

ARYL SULPHATASES

These enzymes occur mainly in animals and microorganisms and are, at present, of no great interest to workers in the plant phenolics field as naturally occurring phenolic sulphates such as persicarin[2] and persicarin-7-methyl ether[3] are quite rare. Much of the work on the sulphatases has been carried out by Dodgson and Spencer and their associates.[4] There is little reason to suppose that the biosynthesis of sulphates occurs by a reversal of aryl sulphatase activity. Activated sulphate derivatives such as adenosine-3′-phosphate-5′-sulphatophosphate (PAPS)[5] are more likely to be involved in such reactions *in vivo*.

ESTERASES

The true esterases, i.e. those which hydrolyse simple esters are common in animal tissues but are infrequently found in higher plants, although wheat germ[6] and citrus esterases[7] have been reported. Lipases, the group of esterases which hydrolyse water-insoluble esters, are widespread in plant tissues.

Aspergillus niger and other moulds yield mixtures of esterases known as *tannases*. These preparations have been particularly useful to organic chemists who are interested in the structures of hydrolysable tannins. Crude tannase from *A. niger* will hydrolyse both aliphatic and depside ester linkages but the preparation shows a high specificity towards the acyl moiety of the ester. For example, it will hydrolyse the methyl esters of 3,4,5-trihydroxy-, 3,4-dihydroxy- and 3,5-dihydroxy-benzoic acids but not those of salicylic acid and 2,4-dihydroxy- and 2,5-dihydroxy-benzoic acids.[8,9]

Toth and Barsony[10] claimed to have isolated a specific depsidase from crude tannase by chromatography. Haslam *et al.*[11] found evidence for the existence of a separate aliphatic esterase and a depsidase in tannase but were unable to achieve any separation of these activities. The Sheffield group[11] were, however, able to obtain a tannase preparation free from glycosidases by ion-exchange chromatography. This was used to investigate the structures of gallotannins[12] and it was clearly shown that they had glucose cores esterified with gallic acid residues and not oligosaccharide cores as claimed by White[13] and Grassmann *et al.*[14]

Nierenstein[15] claimed that esters could be synthesized by a reversal of tannase action. Thus he was able to form *m*-digallic acid by the condensation of two molecules of gallic acid. The physiological significance of this reaction is doubtful. Ester linkages are more likely to be synthesized *in vivo* via activated acyl derivatives. Some fungi can utilize tannin as a carbon source and here the presence of a tannase is obviously important. Tannases have so far not been reported in higher plants.

GLYCOSIDASES

The discussion with this group of enzymes will be chiefly limited to the simple glycosidases, i.e. those which hydrolyse alkyl and aryl glycosides

$$\text{Glycosyl—OR} + \text{EH} \rightleftarrows \text{Glycosyl—E} + \text{R—OH}$$

$$\text{Glycosyl—E} + \text{HO—R}' \rightleftarrows \text{Glycosyl—OR}' + \text{EH}$$

FIG. 1.

and low molecular weight oligosaccharides. The enzymes which hydrolyse polysaccharides and those which are involved in transfer reactions with nucleotide derivatives of sugars, such as uridine diphosphate glucose (UDPG) may also be regarded as glycosidases. These, however, will not be dealt with in any detail.

The enzymic hydrolysis of glycosides occurs by fission of the bond between C—1 of the glycone and the glycosidic oxygen atom.[16] The overall reaction is shown in Fig. 1. R in the glycosyl donor molecule may be an alkyl, aryl or glycosyl group. If R′ in the acceptor molecule is H then normal hydrolytic cleavage of the glycoside occurs with the liberation of free reducing sugar. Where R′ is an alkyl or glycosyl group a new glycoside bond is formed by transglycosylation. Most glycosidases exhibit competing hydrolytic and transferase activities, hydrolysis really being a special form of transglycosylation with water acting as the acceptor molecule.

Hydrolase Activity

The existence of glycoside-hydrolysing enzymes has been recognized for 150 years although a great many questions remain unanswered regarding their mode of action. This is mainly because they frequently occur in complex mixtures which are difficult to fractionate. At the present time only one crystalline glycosidase is known, namely the β-galactosidase from *E. coli*.[17]

The most common sources of glycosidases are the *Rosaceous* plants and in particular sweet almond seeds from which almond emulsin is obtained. The properties of this complex mixture of enzymic proteins has been extensively reviewed.[18–22] The main enzymes present in the emulsin are β-D-glucopyranosidase, β-D-galactopyranosidase, α-D-galactopyranosidase, and α-D-mannopyranosidase. There is some controversy regarding the β-glucosidase and β-galactosidase. Helferich [23] believed that a single enzyme was responsible for both activities. This was disputed[19] but recent evidence[24] again suggests that one active site is responsible for both activities. Almond emulsin also contains minor β-D-glucopyranuronosidase and 2-acetamido-2-deoxy-β-D-glucopyranosidase components, and will hydrolyse α- and β-L-arabinopyranosides, β-D-xylopyranosides and α-D-lyxopyranosides. It is not necessary to postulate the existence of separate enzymes in the case of the pentosides as these may well be hydrolysed by the major hexosidase components with the corresponding structural specificities.

In general terms the specificity of a glycosidase is directed towards four structures in the substrate:

1. Configuration at the anomeric carbon atom.
2. Shape of the sugar ring.
3. Arrangement and substitution of the hydroxyl groups on the sugar ring.
4. Structure of the aglycone.

Aglycone specificity is normally much less stringent than glycone specificity.

In the case of purified almond β-glucopyranosidase the enzyme will not hydrolyse α-glucopyranosides, β-glucofuranosides or β-thioglucopyranosides. Absolute specificity for the configuration of the groups on C—2, C—3, C—4 and C—5 exists and the enzyme will not tolerate substitution of the hydroxyl groups on C—2, C—3 or C—4. Substitutions at C—5 or C—6 affect the rate of hydrolysis but do not normally result in complete inhibition. In general, the larger the substituent at C—6 the slower the rate of hydrolysis. β-Glucopyranosidase has a low degree of specificity with regard to the aglycone although differences in structure do affect the rate of reaction. For example, aryl-β-glucopyranosides are usually hydrolysed more rapidly than alkyl-β-glucopyranosides and substitution of the aglycone moiety in an aryl-β-D-glucopyranoside with an electron-attracting group facilitates enzymic hydrolysis.[25] With other glucopyranosidases there are wide variations in aglycone specificities. Thus the α-D-glucopyranosidase from *Candida tropicalis*[26] will hydrolyse starch, maltose, sucrose, methyl-α-D-glucopyranoside and phenyl-α-D-glucopyranoside. *Stachybotyris atra* β-D-glucopyranosidase[27] on the other hand exhibits a high affinity for aryl-β-D-glucopyranosides.

Transferase Activity

Transglycosylation with the formation of new glycosidic bonds occurs when the acceptor molecule is an alcohol. This was first observed by

FIG. 2.

Rabaté[28] who showed that an emulsin from *Salix purpurea* would transfer glucose from the phenolic hydroxyl group of saligenin to the primary hydroxyl group (Fig. 2), and also that glucose could be transferred from phenyl-β-D-glucopyranoside to primary alcohols. Since these original observations transglycosylation reactions involving many glycosidases have been studied in great detail. This literature has recently been reviewed by Jermyn[29] and Dedonder.[30] It is interesting to note that there is always retention of configuration at C—1 of the transferred glycone moiety in these reactions.*

Glycosidases show some specificity towards acceptor molecules. Courtois and Leclerc,[32] for example, have shown that almond β-glucopyranosidase will transfer glucose from phenyl-β-D-glucopyranoside to

* (In the case of transfer from UDPG inversion at C—1 of the glucosyl residue may occur).[31]

various sugars, simple primary and secondary alcohols but not tertiary alcohols.

Miwa and other Japanese workers using a number of different glycosidases, have shown that each enzyme exibits characteristic affinities towards different acceptor molecules. This work has been summarized by Jermyn.[27]

Attempts to glucosylate phenolic hydroxyl groups using higher plant glycosidases and "low energy" potential glucose donors have failed.[35–36] In other words, phenols would appear to be rather poor acceptor molecules. Pridham,[37] however, has been able to transfer glucose from maltose to resorcinol with the formation of *m*-hydroxyphenyl-α-D-glucopyranoside and *m*-hydroxyphenyl-α-isomaltoside using an *A. niger* enzyme preparation. Other di- and tri-hydric phenols also acted as acceptors in this system. It is at present uncertain whether the enzyme responsible for this reaction falls into the class of simple glycosidases as defined on page 74. At least no dialysable co-factor appeared to be involved in the reaction and transfer from UDPG to resorcinol could not be demonstrated.

Mechanism of Glycosidase Action

The glycosidases are related in a number of ways. They are usually acidic proteins and have no prosthetic groups or coenzymes. Maximum hydrolytic and transferase activities normally occur in neutral or weakly acidic solutions and hydrolysis proceeds by hexose—0 bond fission.[16] With regard to the active sites on these enzymes, few have been examined in detail, but a variety of acidic and basic groups have been implicated. At the present time there is widespread belief that the majority of simple glycosidases have a similar mechanism of action and probably function as acid-base catalysts. In other words, the glycoside undergoes attack by nucleophilic and electrophilic groups on the enzyme surface which results in electron displacement and rupture of the glycosidic bond.

A recent, attractive mechanism for *E. coli* β-galactopyranosidase activity, based on Koshland's double displacement theory[38] has been proposed by Wallenfels and Malhotra.[39] This enzyme is a typical glycosidase and pH-activity and inhibitor studies indicated that sulphydryl and imidazole groups were present on the active site. It is suggested that the enzyme-substrate complex is formed as shown in Fig. 3, the galactosyl residue being held onto the enzyme surface by hydrogen bonds between carbonyl groups of alternate peptide bonds and hydroxyl groups on the sugar. Six to seven amino acid residues are probably present between the cysteine and the histidine. Hydrolysis of the galactoside is effected by nucleophilic attack by the imidazole nitrogen on C—1 of the sugar and simultaneous hydrogen bond formation between the glycosidic oxygen atom and the sulphydryl group (Fig. 4). Bond fission occurs and free galactose or a new galactoside is then produced by a second nucleophilic attack by the acceptor molecule (water or an alcohol). It should be noted that with this mechanism the aglycone group and the acceptor molecule occupy the same position on the enzyme surface. The two S_{N2} reactions produce a double inversion at C—1 and the β-configuration of the original galactoside is thus retained.

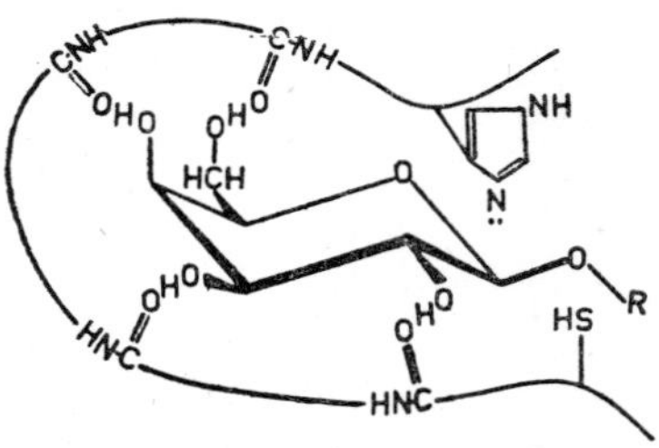

FIG. 3. A possible structure of the β-galactosidase-substrate complex.[39]

With reference to this mechanism and assuming that other glycosidases function in a similar manner, it is interesting to speculate why phenols cannot generally act as acceptor molecules in glycosylation reactions. The answer may of course be that the phenol and the glycosyl-enzyme complex are sterically incompatible. If, however, aglycones and acceptors occupy

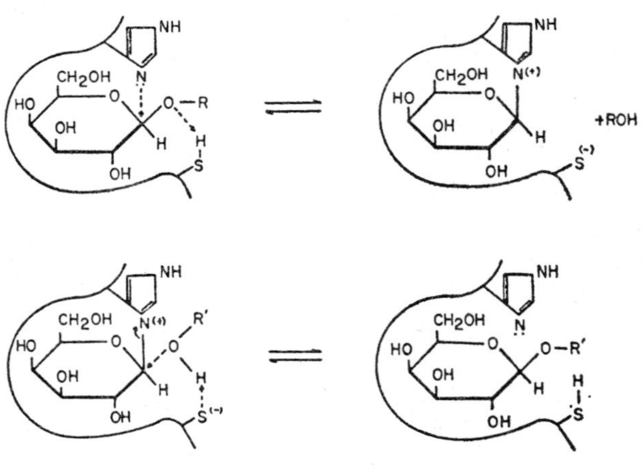

FIG. 4. Proposed mechanism for β-galactosidase action.[39]

the same position on the enzyme surface this argument may not be valid. A second possibility is based on the fact that the electron density at the oxygen atom of a phenolic hydroxyl group is lower than that of an oxygen in an alcoholic hydroxyl group. In other words, if the phenol is a weak nucleophile the formation of the new glycoside bond may be hindered.

PHYSIOLOGICAL IMPORTANCE OF GLYCOSIDASES

It is normally assumed that plant glycosidases can and do hydrolyse glycosides *in vivo*. For example, Freudenberg *et al.*[40] believe that in lignin biosynthesis the first stage is the liberation of coniferyl alcohol from coniferin by a β-glucopyranosidase. Glycosidically bound sugars may also function as energy reserves.[41] In this case one might again assume that glycosidases liberate these sugars before they are metabolized by the plant. The evidence which supports glycosidase activity *in vivo* is, however, only circumstantial and it is not known with any certainty if glycosidases and glycosides ever come into contact in normal healthy tissues. Where cells are mechanically damaged or diseased, contact is certain and subsequent hydrolysis with the liberation of antimicrobic aglycones could in these instances be an important function of glycosidases. The exact significance of other carbohydrases, such as the amylases and invertase, in plant tissues is also not clear.

The participation of simple glycosidases in the biosynthesis of glycosidic bonds in living plant tissues seems very unlikely. The equilibrium constants for reactions of the type shown in Fig. 1, normally favour hydrolysis and not synthesis. In the presence of high substrate concentration, however, a simple transfer reaction could result in glycoside synthesis. Thus 1^{F}-β-fructosylsucrose might well arise by the action of sucrase on sucrose in the cambial tissues of aspen.[34] In the case of the phenolic glycosides it is almost certain that biosynthesis in higher plants is effected via "high energy"[35,42] sugar derivatives such as UDPG. It is possible *in vitro* to synthesize a diglucoside from a monoglucoside using a simple glycosidase (e.g. to convert arbutin to *p*-hydroxyphenyl-β-gentiobioside with almond emulsin[36]) but, here again, *in vivo* the second glucose residue is probably derived from UDPG.[34,43]

REFERENCES

1. E. Racker, in *The Enzymes*, 2nd ed., Vol. 5, p. 313, eds. P. D. Boyer, H. Lardy and K. Myrbäck, Academic Press, New York (1961).
2. H. Tatsuta and Y. Ochii, *J. Chem. Soc. Japan* **75**, 241 (1954).
3. H. Tatsuta, *J. Chem. Soc. Japan* **75**, 939 (1954).
4. See references in *Biochemists' Handbook*, p. 270, ed. C. Long, Spon, London (1961).
5. J. Baddiley and J. G. Buchanan, *Quart. Revs.* **12**, 171 (1958).
6. T. P. Singer and B. H. J. Hofstee, *Arch. Biochem. Biophys.* **18**, 229 (1948).
7. E. F. Jansen, R. Jang and L. R. MacDonnell, *Arch. Biochem. Biophys.* **15**, 415 (1947).
   E. F. Nansen, M-D. F. Nutting and A. K. Balls, *J. Biol. Chem.* **175**, 975 (1948).
8. H. Dyckerhoff and R. Armbruster, *Z. Physiol. Chem.* **219**, 38 (1933).
9. E. Haslam, R. D. Haworth, K. Jones and H. J. Rogers, *J. Chem. Soc.* 1829 (1961).
10. G. Toth and G. Barsony, *Enzymologia* **11**, 19 (1943–45).
11. E. Haslam, R. D. Haworth, S. D. Mills, H. J. Rogers, R. Armitage and T. Searle, *J. Chem. Soc.* 1836 (1961).

12. R. ARMITAGE, G. S. BAYLISS, J. W. GRAMSHAW, E. HASLAM, R. D. HAWORTH, K. JONES, H. J. ROGERS and T. SEARLE, *J. Chem. Soc.* 1842 (1961).
13. T. WHITE, *The Chemistry of Vegetable Tannins*, p. 13, Soc. Leather Trades' Chemists, Croydon (1956).
14. W. GRASSMANN, G. STIEFENHOFER and H. ENDRES, *Ber.* **89,** 454 (1956).
15. M. NIERENSTEIN, *Biochem. J.* **26,** 1093 (1932).
16. C. A. BUNTON, T. A. LEWIS, D. R. LLEWELLYN, H. TRISTRAM and C. A. VERNON, *Nature, Lond.* **174,** 560 (1954).
D. E. KOSHLAND and S. S. STEIN, *J. Biol. Chem.* **208,** 138 (1954).
S. E. SPRINGHORN and D. E. KOSHLAND, 128th Meeting ACS, abstracts, p. 37C (1955).
17. K. WALLENFELS and M. L. ZARNITZ, *Angew Chem.* **69,** 482 (1957).
K. WALLENFELS, M. L. ZARNITZ, G. LAULE, H. BENDER and M. KESER, *Biochem. Z.* **331,** 459 (1959).
A. S. L. HU, R. G. WOLFE and R. J. REITHEL, *Arch. Biochem. Biophys.* **81,** 500 (1959).
18. W. W. PIGMAN, *Adv. Enzymol.* **4,** 41 (1944).
19. A. GOTTSCHALK, *Adv. Carbohydrate Chem.* **5,** 60 (1950).
20. S. VEIBEL, in *The Enzymes*, 1st ed., Vol. 1, Pt. I, pp. 583, 621, eds. J. B. Sumner and K. Myrbäck, Academic Press, New York (1950).
21. H. BAUMANN and W. PIGMAN in *The Carbohydrates*, p. 562, ed. W. Pigman, Academic Press, New York (1957).
22. A. Gottschalk in *Encylopedia of Plant Physiology*, Vol. 6, p. 107, ed. W. Ruhland, Springer-Verlag, Berlin (1958).
23. B. HELFERICH, *Ergeb. Enzymforsch.* **7,** 83 (1938).
24. R. HEYWORTH and P. G. WALKER, *Biochem. J.* **83,** 331 (1962).
25. R. L. NATH and H. RYDON, *Biochem. J.* **57,** 1 (1954).
26. T. SAWAI, *J. Biochem., Tokyo* **44,** 49 (1958).
27. M. A. JERMYN, *Aust. J. Biol. Sci.* **8,** 577 (1955).
28. M. RABATÉ, *Bull. Soc. Chim. Biol.* **17,** 572 (1935).
29. M. A. JERMYN, *Rev. Pure Applied Chem.* **11,** 92 (1961).
30. R. A. DEDONDER, *Ann. Rev. Biochem.* **30,** 347 (1961).
31. W. Z. HASSID, E. F. NEUFELD and D. S. FEINGOLD, *Proc. Natl. Acad. Sci. U.S.* **45,** 905 (1959).
32. J. E. COURTOIS and M. LECLERC, *Bull. Soc. Chim. Biol.* **38,** 365, (1956).
33. C. E. CARDINI and T. YAMAHA, *Nature, Lond.* **182,** 1446 (1958).
34. J. B. PRIDHAM, *Biochem. J.* **76,** 13 (1960).
35. J. B. PRIDHAM and M. J. SALTMARSH, *Biochem. J.* **74,** 42P (1960).
36. J. D. ANDERSON, L. HOUGH and J. B. PRIDHAM, *Biochem. J.* **77,** 564 (1960).
37. J. B. PRIDHAM, *Chem. and Ind.* 1172 (1961).
38. D. E. KOSHLAND in *The Enzymes*, 2nd ed., Vol. 1, p. 305, eds. P. D. Boyer, H. Lardy and K. Myrbäck, Academic Press, New York (1959).
39. K. WALLENFELS and O. P. MALHOTRA in *The Enzymes*, 2nd ed., Vol. 4, p. 409, eds. P. D. Boyer, H. Lardy and K. Myrbäck, Academic Press, New York (1960).
40. K. FREUDENBERG, H. REZNIK, H. ROSENBERG and D. RASENACK, *Ber.* **85,** 641 (1952).
41. W. PFEFFER, *Pflanzen. Physiol.* **1,** 492 (1897).
M. BRIDEL, *Rev. Gén. Sci. Pure Appl.* **37,** 134 (1926).
T. WEEVERS, *Jb. Wiss. Bot.* **39,** 229 (1904).
42. T. YAMAHA and C. E. CARDINI, *Arch. Biochem. Biophys.* **86,** 127 (1960).
43. T. YAMAHA and C. E. CARDINI, *Arch. Biochem. Biophys.* **86,** 133 (1960).

CHAPTER 8

PHENOL OXIDASES IN *STREPTOMYCETES*

E. KÜSTER

Department of Industrial Microbiology, University College, Dublin

SOON after *Streptomycetes* were discovered, their ability to brown organic media was observed and extensively studied. Beijerinck[1] named such strains chromogenous and suggested that this phenomenon was related to a tyrosinase effect. Skinner[2] also called the browning of peptone-containing media with melanin formation a tyrosinase reaction, but an isolation and exact determination of this phenol oxidase has not yet been achieved. *Streptomyces scabies*, the cause of potato scab, was very suitable for examining this melanin production in detail[3] because it was suggested that a relationship between tyrosinase activity and virulence existed.[4] However, this relationship could not be confirmed using genetically changed, but still virulent, strains.[5]

The methods for proving and isolating phenolases which have been successfully employed with higher plants and fungi could not be applied to *Streptomycetes* without further modification. The Bavendamm test,[6] i.e. blackening of a tannic acid medium, is quite useful for the examination of fungi, but is not applicable to *Streptomycetes* because of the strongly acidic conditions. A positive tannic acid reaction indicates the presence of an extracellular polyphenolase of the laccase type.[7] The Bavendamm test has therefore been modified in such a manner that, instead of tannic acid, various oxidizable polyphenols were added to a suitable agar medium. Small agar discs containing grown *Streptomycetes* were put on these agar plates. Tyrosine, *p*-cresol, catechol, quinol, guaiacol, *p*-phenylendiamine and pyrogallol were used as substrates in these tests. After 24 hr in the positive cases, brown or black coloured zones around the discs could be observed.[8] This rough qualitative method facilitates the selection of strains with an enzymic activity and also indicates the presence of different types of phenolases according to their substrate specificities. Quantitative figures were obtained by manometric measurements with the Warburg apparatus. In these experiments it was curious that the consumption of oxygen was lower after the addition of catechol than without it although the reaction solution darkened.[9] It was not clear from these results whether the slight oxygen uptake observed on addition of catechol was due to the complete inhibition of normal endogenous respiration and oxidation of this substance by phenolases or to its partial inhibition. Using malachite green, which inhibits normal endogenous respiration but not the activity of phenolases, it was found that this oxygen uptake was derived from the enzymic oxidation of catechol only. It has also been shown that malachite green specifically inhibits the tyrosinase effect and not the effect of polyphenolases. Surface-active substances, such as sodium cholate and Tween-80 also suppress the endogenous respiration so that the phenol oxidase activity can be determined as reported by Sevcik.[10,11] Colorimetric methods can also be successfully used for the examination of the polyphenolases of *Streptomycetes.* The method of Braesch[12] with *p*-aminobenzoic acid and that of Keilin and Mann[13] which is based on the enzymatic oxidation of pyrogallol to purpurogallin gave reproducible and comparable results. We modified and standardized the purpurogallin method so that significant figures could be obtained with low enzyme activities; the Braesch method needs higher enzyme activities. It is interesting to note that nitrite as a strong oxidizing agent gives an enzyme-like effect when it is formed from nitrate by nitrate-reducing *Streptomycetes.*[14]

All the above methods have been applied to show that phenolases are present in the *Streptomycetes.* If we know the distribution of these enzymes the question then arises whether there are specific functions which distinguish the *Streptomycetes* from other organisms. Phenolases as oxidizing enzymes oxidize phenolic compounds with the formation of quinones. It is known that phenolases in some cases may play a role as terminal oxidases for respiration in so far as the oxygen uptake by cytochrome oxidase is taken over by phenolase, e.g. in fly larvae.[15] However, this is not true for the *Streptomycetes*[16] or for some fungi.[17] Frommer[16] concluded that pigments and antibiotics with quinonoid structures most probably do not have a respiratory function *in vivo*; it is more likely that they are by-products of the organism's metabolic activity, produced when excess intermediates accumulate. Frommer's ideas suggest that these substances are formed and secreted in order to detoxicate and protect the organisms themselves, rather than as antibiotics against other organisms. Interesting studies have been carried out which demonstrate a relationship between the occurrence and activity of phenolases and the formation of antibiotics. Frommer[18] examined actinomycin- and rhodomycin-producing *Streptomycetes* and showed the presence of phenolases. Phenylthiourea, as an inhibitor of tyrosinase, suppresses the formation of these pigmented antibiotics.[16] Sevcik[19,20] found that a direct relationship existed between the amount of actinomycin produced and the activity of phenolases. He concluded that phenolases of the laccase type may play a role in the biosynthesis of actinomycin by *Streptomyces antibioticus.* 2,5-Dihydroxytoluquinone for instance has been found as a degradation product of actinomycin.[21]

We carried out similar experiments with *Streptomyces coelicolor* which forms a blue–red pigment which behaves as a pH indicator and also possesses a quinonoid constitution.(22) The production of the pigment by some strains can be directly related to increased phenolase activity and increase in the pH of the medium (Table 1).(14)

It is suggested that one of the possible pathways of humic acid formation involves quinonoid precursors. Therefore a relationship between phenolase activity and humic acid formation should also be demonstrable. In fact this has been proved (Table 2).(9) Several different strains of *Streptomycetes* were cultivated on glycerol–glycine-solutions for 3 months. The humus-like substances were then precipitated with HCl and extracted with 1-butanol. The butanol-soluble material was quantitatively measured as

TABLE 1. PIGMENT FORMATION BY *Streptomyces coelicolor* AND CATECHOL OXIDATION

Strain No.	Age in days	pH of filtrate	Pigment (mg/100 ml filtrate)	O_2 uptake with malachite green and catechol (μl/100 ml filtrate)
K473	13	5·4	0·0	17·4
K472	14	5·1	22·4	56·8
K808	16	6·5	26·6	74·4
G434	20	7·1	41·8	82·0

"melanin" in a spectrophotometer. The butanol-insoluble proportion ("humic acids") was dried and weighed. Strain B33 gave the lowest yields of "melanin" and "humic acids" and the other yields listed in Table 2 are therefore quoted relative to results obtained with B33. The same strains have been tested by the plate method with quinol as substrate. Quinol is oxidized by laccase but not by tyrosinase; increase in the zones on the quinol plates corresponds to an increase in humus formation. In addition, the oxygen uptake of extracts of these 3 months' old mycelia was measured manometrically with tyrosine and catechol as substrates. In both cases the values for the strains B47 and B63 are twice those of the other strains. The manometric measurements with strains B17 and B63 indicate that the activity of polyphenolases increases at the same rate as the "humic acid" formation, whereas the tyrosinase activity is proportional to the "melanin" formation. It can be seen that the presence and activity of polyphenolases, probably of the laccase type, are important factors for humic acid formation. This is true in so far as quinonoid metabolites are produced which

TABLE 2. FORMATION OF HUMIC ACIDS AND PHENOLASE ACTIVITY IN *Streptomycetes*

Strain no.	Colour of filtrate	Humic acid (mg/1200 ml filtrate)	Relative amount of humic acid (B33 taken as 1)	Relative extinction of 1-butanol extract (B33 taken as 1)	Zones on quinol plates (mm)	O_2 uptake (μl/10 mg mycelium)	
						Tyrosine	Catechol+ Malachite green
B33	Brown	8·4	1	1	3	—	—
H44	Brown	29·8	3·5	3	2–3	—	—
W40	Dark red–Brown	53·6	6·4	5	4	—	—
B17	Brown–black	56·0	6·7	100	3	4	1
B47	Brown–black	93·0	11·1	100	6–7	—	—
B63	Black	110·0	13·1	100	6–7	4·1	2·6

act as humic acid precursors. How far phenolases also take part in further polymerizing reactions by dehydrogenation is not yet known. The following observation suggests that these processes of humic acid formation with regard to enzyme activity occur not only in laboratory model experiments but also to a certain extent in soil under natural conditions. It is well known that the formation of stable humus is greatest under grassland. The number of peptone-browning chromogenous *Streptomycetes* isolated from grassland is generally higher than from arable land of the same soil type.[23] If we consider the peptone-browning, i.e. melanin formation, as a parallel or an intermediate step in humus formation we may perhaps conclude that organisms with phenolases are important and essential for the humus formation in soil.

It has been demonstrated in extensive studies that the peptone-browning reaction qualitatively and quantitatively depends on the presence and activity of phenolases.[24] Also, it was discovered that an addition of amino acids, which do not serve as oxidizable substrates for phenolases, stimulate the browning reaction. Fåhraeus *et al.*[25,26] obtained similar results from experiments with the fungus *Polyporus versicolor* in which tyrosine and 2,5-xylidine induce the formation and activity of laccase. This means that the organism itself forms aromatic compounds which are oxidized by the phenolase present. This also confirms the hypothesis that humus formation takes place with aromatic compounds as intermediates.

In previous studies[8,14] it was suggested that a relationship exists between the Bavendamm reaction and the formation of humic acids. It was found that *Aspergillus fumigatus*, which produces fumigatin (3-hydroxy-4-methoxy-2,5-toluquinone), which can tautomerize to give the *o*-quinone, forms humic acids in autolysed cultures, while *Penicillium spinulosum* which forms spinulosin (6-hydroxy-fumigatin), a quinone with a stable structure, does not yield humic acids. It was also found that with metabolic products of other fungi and *Streptomycetes* only *o*-quinones or quinones which readily change to the ortho form could act as humic acid precursors. Recent enzymic studies[27] on fumigatin and spinulosin have shown that the formation of both these substances is due to the *p*-phenolase activity of laccase.[28] These enzymes were found to occur endo- and exo-cellularly in the case of fumigatin formation and only endo-cellularly when spinulosin was formed. *Aspergillus fumigatus* gives a positive Bavendamm reaction, while *Penicillium spinulosum* is negative. It was shown that this reaction is not quantitatively related to fumigatin formation and is therefore due to exocellular phenolases only. Hence, humic acid formation appears to depend on the nature of the quinonoid precursors and not on the enzymes forming them. The functions of the exocellular phenolases has not yet been established. At present it is not known if the same enzyme relationships exist for the *Streptomycetes* but the laccase which Sevcik[11] has found to be responsible for the formation of actinomycin is an endo-enzyme. Also, the phenolases which have been considered in relation to melanin and humus formation are, at least partially, endo-enzymes. The endo- and exo-cellular character of the phenolases has to be considered before any general conclusions with regard to their function can be made.

SUMMARY

Methods have been developed which show that phenolases are frequently present in *Streptomycetes*. These phenolases oxidize phenolic compounds which are either formed by the organism itself or supplied from the environment. The differentiation between and the importance of exo- and endo-cellular phenolases is discussed.

Quinonoid substances formed by phenolases are constituents of some antibiotics. Likewise, quinones and in particular *o*-quinones act as precursors of humic acids. Polyphenolases of the laccase type are responsible for the formation of these quinones. Relationships between the presence and activity of phenolases and the formation of antibiotics and humic acids have been observed.

REFERENCES

1. M. W. BEIJERINCK, *Ztrlbl. Bakt.* II Abt. **6,** 2 (1900).
2. C. E. SKINNER, *J. Bact.* **35,** 415 (1938).
3. J. P. HOLLIS, *Phytopathology* **43,** 355 (1953).
4. E. B. VAISEY, J. A. CARPENTER and R. G. ATKINSON, *Canad. J. Microbiol.* **1,** 574 (1955).
5. K. F. GREGORY and E. B. VAISEY, *Canad. J. Micobiol.* **2,** 65 (1956).
6. W. BAVENDAMM, *Z. Pflzkrankh. Pflzschutz.* **38,** 257 (1928).
7. H. LYR, *Planta* **50,** 359 (1958).
8. E. KÜSTER, *Ztrlbl. Bakt. I Orig.* **160,** 207 (1953).
9. E. KÜSTER, *Z. Pflanzenern. Düngung Bodenkd.* **69,** 137 (1955).
10. V. SEVCIK, *Nature, Lond.* **179,** 966 (1957).
11. V. SEVCIK, *Folio Biol.* **3,** 212 (1957).
12. S. BRAESCH, *Bull. Soc. Chim. Biol.* **36,** 711 (1954).
13. D. KEILIN and T. MANN, *Proc. Roy. Soc. London* Ser. B **125,** 187 (1938).
14. E. KÜSTER, *Giorn. Microbiol.* **2,** 351 (1956).
15. P. KARLSON and E. WECKER, *Z. Physiol. Chem.* **300,** 43 (1955).
16. W. FROMMER, *Z. Physiol. Chem.* **307,** 124 (1957).
17. D. BOULTER, *J. Gen. Microbiol.* **16,** 305 (1957).
18. W. FROMMER, *Arch. Mikrobiol.* **23,** 385 (1956).
19. W. SEVCIK, *Nature, Lond.* **179,** 1191 (1957).
20. V. SEVCIK, *Folia Biol.* **4,** 156 (1958).
21. H. BROCKMANN and H. MUXFELDT, *Angew. Chem.* **68,** 67 (1956).
22. H. BROCKMANN, W. MÜLLER and K. VAN DER MERVE, *Naturwiss.* **49,** 131 (1962).
23. W. FLAIG and H. J. KUTZNER, *Arch. Mikrobiol.* **35,** 207 (1960).
24. E. KÜSTER, *Ztrlbl. Bakt.* II. Abt. **111,** 227 (1958).
25. G. FÅHRAEUS and G. LINDEBERG, *Physiol. Plant.* **6,** 150 (1953).
26. G. FÅHRAEUS, V. TULLANDER and H. LJUNGGREN, *Physiol. Plant.* **11,** 631 (1958).
27. E. KÜSTER and B. T. LITTLE, *Biochim. Biophys. Acta* in press.
28. G. FÅHRAEUS and H. LJUNGGREN, *Biochim. Biophys. Acta* **46,** 22 (1961).

CHAPTER 9

ENZYME INHIBITION BY PHENOLIC COMPOUNDS

A. H. WILLIAMS

Long Ashton Research Station, The University, Bristol

THIS paper does not attempt a complete survey of the subject, but discusses certain aspects with which the author has some direct connection.

At this symposium much has been said about the action of enzymes on phenolic compounds, and in a sense this paper simply looks at the same process from a different viewpoint. It is, however, often helpful to look at the process of mutual interaction of enzyme and phenolic from one side or the other; if after interaction, the phenolic is largely changed while the enzyme retains its activity, the enzyme can be said to act on the phenolic; but if, on the other hand, the enzyme is rapidly deactivated and most of the phenolic remains unchanged, the phenolic would be said to act on, or inhibit, the enzyme. This simple picture is complicated by a third case in which the enzyme acts on the phenolic at first but is then deactivated by the products of its own action.

Various aspects of the first case have been discussed in several of the other papers presented at this Symposium. The second case has fascinated many botanists, as the many references in the literature to the protective function of plant phenolics bear witness. They are often claimed to confer resistance to infection by fungi or viruses, presumably by inactivation of the fungal enzymes or the virus nucleoprotein The evidence for many of these claims is slight; the idea may even derive in part from analogy with the medical use of phenols as antiseptics. The resistance of the plant may be real enough, but its dependence on the phenolic content seems often to be doubtful. For example, to confer resistance to virus infection, a phenolic would be expected to combine with the virus, inactivating its nucleoproteins. This is the explanation usually put forward for the difficulty, or impossibility, of the mechanical transmission of virus in the rosaceous fruit plants; but when the idea is considered in detail there are some serious difficulties. If the virus is to be inactivated outside the cell, that is to say during transmission, the phenolic must be present in the extra-cellular spaces (and there is little evidence in favour of this) or be liberated when the cells are damaged. Since virus can proliferate in undamaged cells, either the phenolics in the cell do not inactivate the virus, or else they are localized even within the cell and kept separate from the virus, so that only when the cell is damaged can interaction take place. But while mechanical transmission of virus in the apple tree is not possible in the sense of rubbing together abraded surfaces of infected and healthy plants, transmission by budding is quite possible even if the bud is only in position for an hour or two; and budding involves quite considerable mechanical damage to tissues, which would be expected to liberate localized phenolics. The simplest explanation is that the phenolics must be modified before they can function an inhibitors, most probably by oxidation; ordinary mechanical transmission involves exposing damaged surfaces freely to the air, while in budding it is possible that the region between the bud shield and the host cambium will have areas free of air in which non-oxidizing conditions are re-established with reversal of the initial oxidation of *o*-dihydroxy groups to quinones. Visual evidence that this sort of reversal can take place is seen in milled apple fruit tissue; the initial browning proceeds further on the exposed surface but disappears in the underlying mass away from air so long as this still contains unoxidized ascorbic acid.

It may be, then, that the simple phenolics really have very little inhibitory effect on enzymes, and that more complex molecules formed by oxidative coupling or natural polymerization are really necessary; in fact, that the action is largely by the tanning or precipitation of the protein part of the enzyme, a process analogous to that discussed by Cadman[1] in relation to the inactivation of raspberry virus.

The essential property for this sort of reaction is a certain minimum molecular size of the "tannin" molecule, and this very property makes it unable to penetrate cell walls and move about within the plant.

Flood and Kirkham[2] in their work on *Venturia* infections of apple and pear found in their early experiments that simple phenolics such as arbutin and chlorogenic acid, which occur in the leaf of the pear, had an inhibitory action on the growth and sporulation of the appropriate fungus and could move in the plant from the point of injection; but at the same time anomalous results were obtained, such as the stimulation of sporulation of *V. pirina* by phloridzin and quinol. Further investigations with related compounds on the variation of hydroxylation pattern is substituted cinnamic acids indicated that the most active inhibitors were a "foreign" phenolic, *o*-coumaric acid, and a non-phenolic substance, cinnamic acid itself. These results hardly support the original idea of the unique importance of the simple phenolic substances of the plant in protection against *Venturia* infections, and even the effects observed were greatly dependent on the nutritional status of the plant with regard to nitrogen.

At Long Ashton, Dr. Byrde and I have tested the effects of phenolics on the germination of spores of *Sclerotinia fructigena* (fungal Brown Rot of fruit). The results with a selection of simple phenols and their derivatives are shown in Table 1.

Substances with more than two phenolic hydroxyls in the molecule were rather ineffective, and completely methylated compounds also showed little activity, though this may be due to their very limited solubility in water. Activity was greatest for certain monophenols.

TABLE 1. INHIBITION (%) OF SPORE GERMINATION OF *Sclerotinia fructigena* BY PHENOLIC AND RELATED COMPOUNDS (0·01 M OR SATURATED SOLUTION AT pH 4) CORRECTED FOR CONTROL VIABILITY; BASED ON COUNT OF 100 SPORES

Catechol	89	Catechol monomethyl ether	14	Catechol dimethyl ether	6
Resorcinol	70	Resorcinol monomethyl ether	100	Resorcinol dimethyl ether	6
Quinol	21	Quinol monomethyl ether	100	Quinol dimethyl ether	0
Pyrogallol	49	Phloridzin	0		
Phloroglucinol	45	Catechin	14		
Thymol	100	Epicatechin	17		

An examination of the inhibition by phenolic and related acids (Table 2) shows that a very common natural phenolic, chlorogenic acid, and its components, caffeic and quinic acids, are exceedingly ineffective. Monophenolic acids are rather more active, while the best results are from methylated or non-phenolic aromatic acids. Hydroxylation seems, in fact, to be a disadvantage, as is well illustrated in the series of mono-, di- and tri-hydroxy benzoic acids.

These results are almost certainly a reflection of ability to penetrate cell membranes as well as of innate toxicity, but they do cast doubt on the idea that phenolic substances have by virtue of their phenolic nature a direct and simple protective action in plants against infection by pathogens.

In spite of the apparent ineffectiveness of phenolics against the germination of *Sclerotinia* there remains the fact that cider apples with a high phenolic content are more resistant to infection by this fungus through wounds than are dessert varieties. The resistance is only to infection; once established, the fungus spreads through the fruit apparently quite regardless of the phenolic content. Some varieties do in fact show more resistance than others to the spread of the fungus, but they are low-tannin culinary varieties such as Bramley and Edward VII.[(3)] Resistance to *wound* infection was observed to be associated with the rate of browning of the injured tissue surface,[(4)] which is dependent both on the phenolic content and the activity of the phenol oxidase enzymes in the fruit tissue, which in itself is a function of both the inherent activity of the enzyme and the pH of the fruit.

Spores of *S. fructigena* germinate quite normally in the oxidized juice of cider varieties; evidently the oxidized phenolics have as little fungistatic activity as the unoxidized compounds; but the spread of infection from a spore-inoculated "browned" wound is much less than from an unbrowned wound, whether in a variety that shows little browning or in a rapidly browning variety in which browning of the wound has been prevented by painting with glutathione solution.

The key to the problem was the observation of Cole[5] that the pectolytic enzymes of *S. fructigena* were inhibited by apple juice. An examination of the effect showed the juice of high-tannin cider varieties to be particularly

TABLE 2. INHIBITION (%) OF SPORE GERMINATION OF *S. fructigena* BY PHENOLIC AND RELATED ACIDS

Chlorogenic acid	1				
Caffeic acid	8	Quinic acid	0		
p-Coumaric acid	55	*p*-methoxy cinnamic acid	99		
m-Coumaric acid	87	*m*-methoxy cinnamic acid	99		
o-Coumaric acid	83	*o*-methoxy cinnamic acid	97		
Cinnamic acid	100				
Hydrocinnamic acid	100	*p*-hydroxy hydrocinnamic acid	19	3,4-dihydroxy hydrocinnamic acid	13
Phenylacetic acid	99	*p*-hydroxy phenylacetic acid	0	3,4-dihydroxy phenylacetic acid	13
Benzoic acid	100	*o*-hydroxy benzoic acid	100	*o*-methoxy benzoic acid	100
		m-hydroxy benzoic acid	27	*m*-methoxy benzoic acid	100
		p-hydroxy benzoic acid	0	*p*-methoxy benzoic acid	100
Dihydroxybenzoic acids and derivatives		2,3-	81	3,5-	6
		2,4-	79	3,4-	10
		2,5-	5	vanillic acid	99
		2,6-	0	veratric acid	97
Trihydroxybenzoic acids		2,3,4-	8		
		3,4,5- (gallic acid)	0		

effective. Both macerating and endo-PG (polygalacturonase) activities are reduced. The effect is on the external, secreted pectolytic enzymes, for as previously mentioned germination of the spores is not inhibited by browning; but in a browned wound the secreted pectolytic enzymes of the fungus, on which it relies to penetrate the host's cell walls, are inactivated by the complex "tannins" formed during browning. The effect of browned juice on the secreted enzymes can be reduced by treatment with gelatine, known to remove complex tannins by precipitation.

We were interested in trying to sort out which of the phenolics known to be present in apple fruit contributed to this effect. Cole[6] has shown that the leuco-anthocyanins are the most effective, especially when oxidized. We tested individual phenolics known to be present in the apple, before and after oxidation.[7] Oxidized leuco-anthocyanin Ll was found to be the most effective; oxidized catechins were quite effective, chlorogenic acid much less so. (Table 3).

More recent work by Cole and Wood[8] on the reaction between secreted endo-PG enzymes and various fractions of apple phenolics obtained by a rather arbitrary separation confirmed that inhibition was due mainly to complex leuco-anthocyanin material. Fractions containing leuco-anthocyanins of low R_f value in *n*-butanol–acetic acid–water solvent (i.e. of large molecular size), from unoxidized juice, gave over 40 per cent inhibition of endo-PG activity compared with controls, even in the presence of cyanide

TABLE 3. EFFECT OF PHENOLICS ON TISSUE-MACERATING AND ENDO-PG ACTIVITY OF CULTURE FILTRATES OF *Sclerotinia fructigena* AT pH 5. 0·01 M SOLUTIONS UNOXIDIZED (U) OR ENZYMICALLY OXIDIZED (E)

		% reduction in activity	
		Maceration	Endo-PG (agar plate) method
Catechol	U	0	18
	E	67	63
Chlorogenic acid	U	12	3
	E	34	11
Catechin	U	0	19
	E	67	80
Epicatechin	U	0	0
	E	61	93
Leuco-anthocyanin (Cacao Ll)	U	0	11
	E	72	95

which prevents oxidation of the phenolics by polyphenol oxidase from the fungus. In the absence of cyanide, when oxidation takes place, inhibition rose to over 60 per cent. Simpler leuco-anthocyanins (of higher R_f value) gave 10 per cent inhibition in the presence of cyanide rising to over 50 per cent when oxidation, leading to increased molecular size, was permitted. Chlorogenic acid was comparatively inactive even when oxidized. Owing to the composition of the fractions it is difficult to pick out the effect of catechins here.

Since the oxidation products of catechins and leuco-anthocyanins appear to be large, complex molecules we have tried to assess how large a molecule is necessary for the inactivation of pectic enzymes. At an early stage it was shown that tannic acid (which approximates to glucose linked to nine galloyl residues) was an extremely efficient inhibitor of the enzymes. We have since tested a selection of phenolic compounds with medium or large size molecules in their unoxidized state for activity against the external pectolytic enzymes of *S. fructigena*.

Among the gallotannin substances (Table 4) galloyl quinic acids from Tara tannin show some inhibitory activity; smaller molecules have scarcely any, while the larger molecules of the gallotannins are highly active.

TABLE 4. EFFECT OF PHENOLICS ON MACERATING (M) AND POLYGALACTURONASE (ENDO-PG) ACTIVITY OF *S. fructigena*, ENZYMES

	Conc. $\times 10^{-3}$ G-equiv	% reduction in activity	
		M	endo-PG
Chinese gallotannin	1	40	34
	10	63	85
Turkish gallotannin	1	40	28
	10	63	75
Pentagalloyl glucose	1	46	38
	10	70	96
Gallic acid	10	10	7
Digallic acid	1	0	0
Digalloyl quinic acids:			
A	10	28	15
B	10	33	19
C	10	0	13
D	10	58	13
Mixture of tri- and tetra-galloyl quinic acids	10	64	21

Attempts to test other galloyl derivatives such as the myrobolan group and other substances isolated by Professor O. T. Schmidt were limited either by the small amounts of materials available or by their limited solubilities. They could only be tested at low concentrations and under these conditions were relatively inactive with the striking exception of valonic acid dilactone; in view of the comparatively small molecular size and its limited solubility, its activity, which was greater than that of the gallotannins, is most interesting and suggests perhaps some specific effect on the enzyme.

Lastly, we tested a group of tea phenolics the nature of which had been clearly demonstrated by Roberts and Myers[9] (Table 5). The largest of the simpler molecules in this group, epigallocatechin gallate, stands out from the rest; it seems just to cross the threshold of inhibitory activity as its concentration is increased in the unoxidized state; an interesting confirmation, for a different enzyme, of Robert's observation of the inhibitory action of this compound on tea polyphenol oxidase.

The substances of larger molecular size, occurring in manufactured tea are inhibitory, as would be expected, but not notably more so than epigallocatechin gallate itself.

Because of the variables involved in the methods used for estimating enzyme activity, and difficulties imposed by solubility problems, it is difficult to draw firm conclusions from these results; only some tentative suggestions can be made. The minimum molecular size for inhibitory

TABLE 5. EFFECT OF TEA PHENOLICS ON *S. fructigena* ENZYMES

	Conc.	% reduction in activity	
		M*	endo-PG*
Epicatechin	10^{-2}M	0 (61)†	0 (93)
Epigallocatechin	10^{-2}M	0 (14)	0 (26)
Epigallocatechin gallate	10^{-2}M	82 (85)	35 (74)
	10^{-3}M	0	0
Theaflavin	10^{-2}M (0·76%)	56	72
Thearubigen I	0·7%	78	73
Thearubigen II	0·7%	74	62

* Mg macerating enzyme; endo-PG, polygalacturonase.

† Figures in brackets are for the oxidized compounds, illustrating the activity conferred by increase in molecular size.

activity at concentrations up to about 0·2 per cent seem to be in the region of molecular weight 500 and upwards, except for an outstandingly active substance such as valonodilactone which with a molecular weight just below 500 is active at one-tenth of this concentration.

We ought now, perhaps, to look more closely at the nature of the pectic enzymes. Two main types are recognized; pectin (methyl) esterase (PE or PME) which removes methyl alcohol from esterified pectic acids (known as pectinic acids or pectin), and polygalacturonase which splits the basic chains of galacturonic acid units. This second type is often subdivided into those attacking primarily unesterified material (pectic acid), the PG type, and others which more readily attack esterified material (pectin), the PMG type; both of these may again be subdivided into endo-types breaking the polygalacturonic chain at random, and exo-types which split off end groups preferentially.(10) Viewed in this light, our methods for endo-PG do measure this activity, as defined, but our maceration activity is harder to classify since it may include an element of endo-PMG activity;

how much relative to endo-PG it is impossible to say, and one cannot really start drawing conclusions about differential inhibition of PMG or PG activity by individual phenolics. Other work at Long Ashton by Pollard, Kieser and Sissons[11], however, does give some information about this. They examined the effect of pear fruit leuco-anthocyanin, a large and complex molecule, on polygalacturonase activities determined by viscosity measurements. The pear material was strongly inhibitory to yeast polygalacturonase (endo-PG), but two commercial enzyme preparations showed rather more resistance to inhibition of PG, and much more to inhibition of PMG activity. In the same series of experiments it was shown that the activity of apple PME, as measured by methanol production was inhibited by both apple and pear fruit phenolics, and that pre-treatment of the phenolic solution with gelatine reduced the effect. These results suggest that inhibition by phenolics of suitable structure is a feature common to several types of, or perhaps all, pectic enzymes; though their sensitivity does vary. This may, of course, be due to protection of sensitive centres perhaps by other parts of the molecule or by associated impurities. The situation in systems containing many natural products may be exceedingly complicated; thus Ramaswamy and Lamb[12] demonstrated the existence of PME in tea leaf, and showed that during fermentation its activity was reduced by oxidized phenolics made by the action of the tea polyphenol oxidase; this oxidation was, however, impeded by the pectic acid produced by the PME itself. This is a good illustration of the sort of difficulties which may arise in attempts to examine reactions in a system somewhat nearer to *in vivo* in order to extend work beyond the artificiality of *in vitro* tests.

The work of Cole and Wood[8] also illustrates the different results obtained under differing conditions. They showed that only PE activity can be extracted from healthy apple tissue; there is no PG or macerating activity. Tissue rotted by *Sclerotinia fructigena*, which gives brown rots (i.e. the phenolics are oxidized) gives extracts that apparently show enhanced PE activity in spite of the presence of oxidized phenolics. No PG or macerating activity can be extracted, so the fungal external PG enzymes have been inactivated. But from apples rotted by *Penicillium expansum*, which gives a white rot (i.e. the phenolics are not oxidized), high PE, PG and macerating activities can be extracted. Thus here, at first sight, the PG enzymes seem to be inactivated by oxidized phenolics while PE enzymes are unaffected, in contrast to the results of Pollard *et al.*[11] on apple PE, and Ramaswamy and Lamb[12] with tea.

A reconciliation between these sets of results may be possible when the exact conditions of experiment are considered more closely. Cole and Wood measured the PE activity both of juice expressed at about pH 4, and also of salt extracts of the already pressed tissues, using NaCl or Na_2HPO_4 at pH 7. Of the total activity, expressed and extracted, sound fruit gave 23 per cent of its PE activity in expressed juice, but *S. fructigena* rots gave only 2 per cent of their much greater PE activity (fruit+fungal), while *P. expansum* rots gave about 90 per cent in expressed juice. Cole and Wood interpret this in terms of the loss of pectic substances in rots (the greater the loss of these, the less PE activity is fixed) but it may equally well be due to the fixing of PE by oxidized tannins at the low pH of the fruit; at the higher PH of the salt extraction the adsorption complex may be dissociated. This agrees with the observations mentioned by Ramaswamy and Lamb that the optimum extraction of PE from tea leaf occurs at pH 7. There is, however, no corresponding extraction of PG activity from *S. fructigena* rots; the fungal PG seems to be firmly fixed even at pH 7. But with *P. expansum* rots, containing no oxidized phenolics, over 60 per cent of the PG activity can be pressed out in the juice. Thus there may be a considerable difference in the strength of bonding between oxidized phenolics and different pectic enzymes, PG apparently being much more firmly bonded than PE, and the binding is influenced by pH.

These examples show that while inhibition of pectic enzymes by tannins of high molecular weight is fairly well established, the details of the process await further exploration. Other enzyme types such as polyphenol oxidase and cellulase are known to be inhibited by substances of phenolic nature; in these cases too it seems that leuco-anthocyanins and oxidized catechins are implicated, thus supporting the view that it is the more complex phenolics of high molecular weight that are the active components.

Acknowledgements

The work at Long Ashton was carried out in collaboration with Dr. R. J. W. Byrde, to whom the author is indebted for discussion of many points in this paper.

REFERENCES

1. C. H. Cadman in *Phenolics in Plants in Health and Disease*, p. 101, ed. J. B. Pridham, Pergamon Press, Oxford (1960).
2. A. E. Flood and D. S. Kirkham in *Phenolics in Plants in Health and Disease*, p. 81, ed. J. B. Pridham, Pergamon Press, Oxford (1960).
3. R. J. W. Byrde, *J. Hort. Sci.* **31**, 188 (1956).
4. R. J. W. Byrde, *J. Hort. Sci.* **32**, 227 (1957).
5. J. S. Cole, *Ann. Bot.* N.S. **20**, 15 (1956).
6. H. Cole, *Nature, Lond.* **181**, 1596 (1958).
7. R. J. W. Byrde, A. H. Fielding and A. H. Williams in *Phenolics in Plants in Health and Disease*, p. 95, ed. J. B. Pridham, Pergamon Press, Oxford (1960).
8. M. Cole and R. K. S. Wood, *Ann. Bot.* N.S. **25**, 435 (1961).
9. E. A. H. Roberts and M. Myers, *J. Sci. Food. Agric.* **10**, 176 (1959).
10. A. L. Demain and H. J. Pfaff, *Wallerstein Labs. Commun.* **20**, 119 (1957).
11. A. Pollard, M. E. Kieser and D. J. Sissons, *Chem. and Ind.* 952 (1958).
12. M. S. Ramaswamy and J. Lamb, *J. Sci. Food Agric.* **9**, 46, 51 (1958).

CHAPTER 10

TANNIN INHIBITION OF PLANT MITOCHONDRIA

A. C. HULME and J. D. JONES

Ditton Laboratory, Agricultural Research Council, Ditton, Maidstone

OUR INTEREST in the effect of phenolic compounds on the activity of enzymes arose from the difficulty experienced in obtaining active mitochondria from apple peel tissue, and the present paper deals almost exclusively with mitochondrial preparations from this tissue. In addition to its high acidity, apple tissue also contains large amounts of "tannins" especially in very immature fruit.

The main theme of the present work is the effect of various phenolic compounds, monomers as well as polymers corresponding with the classic "tannins", on the Kreb's cycle activity of mitochondria isolated from apple tissues by a method described elsewhere.[1] In the course of the work it became essential to take into account the phenolase activity exhibited by the preparations. Phenolase has often been noted in mitochondrial preparations from plants[2] and its action will obviously influence the overall picture of the effect of the addition of phenolic compounds on the organic acid metabolism of the mitochondria. Our preparations also contain a manganese-dependent enzyme which decarboxylates malate and some results concerning this enzyme will be described.

The mitochondrion is clearly not an ideal subject for a study of the effects of tannins on enzymes as such, since it is an organized system enclosed in a membrane having special permeability properties. We make no apology for this since we feel that not enough attention has been paid to the difference in the natural environment of plant as compared with animal mitochondria. The plant cell is generally much less specialized than the cells of most animal organs and the energy provided through the operation of the Kreb's cycle may be utilized in biochemical processes of quite different types taking place simultaneously in close spatial proximity. If this is so the mitochondria of plants might be expected to have built-in auxiliary systems differing markedly from those associated with mitochondria from animals. Many of the phenolic compounds found in plants (and the apple fruit is a good example of this) have no counterpart in animal organs. The results to be described, being of an exploratory nature, do not give a complete picture of the reaction of mitochondria to the various groups of phenolic compounds. This is partly due to the complexity of the problem and partly due to the difficulty of obtaining some of the compounds we would have liked to study.

It is essential when describing the activity of "mitochondria" to be sure that one is in fact dealing with virtually intact mitochondria and not just particles obtained by the usual routine preparative methods. This is not an easy matter since "essential criteria" for the intactness of mitochondria appear in the literature almost weekly. I will not go in any detail into the reasons why we feel that our mitochondria are virtually intact. They exhibit all the usual biochemical characteristics of integrity including reasonably high ratios for respiratory control, high P/O ratios (examples of these will be seen later) and, through the kindness of Professor Mühlethaler of Zürich, they have been shown to have an appearance in the electron microscope identical with those in intact tissue, complete with double membrane and tubules or cristae.

As a measure of the activity of the mitochondria in inhibition studies we have generally made use of the succinoxidase activity, determinations being made in a Warburg manometric apparatus. Both oxygen uptake and carbon dioxide output were measured since with phenolic compounds the inhibitions may not be specific, decarboxylases as well as oxidase systems being likely to be involved. Determinations of total tannin, leuco-anthocyanin, etc., were made by the methods described by Swain and Hillis.[3]

In our early experiments, the particles we obtained by the usual methods had very little activity even when vacuum infiltration of the tissue with buffer and addition of reducing agents to prevent browning of the tissue during maceration were employed at low temperatures. Another difficulty arose even when some small activity was obtained. On what basis were activities to be expressed for purposes of comparison? Usually, mitochondrial activity is expressed in terms of nitrogen content (generally after treatment with trichloroacetic acid) of the preparation. However, as will be seen later, the nitrogen content of our preparations depended on their tannin content and did not, therefore, give a true measure of mitochondrial protein. Our final method for preparing the mitochondria gave a succinoxidase activity reproducible to about 10 per cent with fruits at comparable stages of development. We have, therefore, found it most satisfactory at present to express activities in terms of the amount of apple tissue (usually 2·5 g) represented by 0·5 ml of mitochondrial suspension. When it is necessary to express results on a nitrogen basis, the total nitrogen content (corrected for any nitrogen compounds added to the mitochondrial suspension) is used in the calculations. The reason for this will be given later.

While the evidence that phenolic compounds inhibit the mitochondrial system has often been indirect (the removal of peel from apples before extraction, the use of reducing substances to prevent browning during extraction, etc.) some direct evidence has been advanced; for example, catechol and chlorogenic acid, almost completely inhibit the major activities of sweet potato mitochondria.(4) Some previous workers have been unable to obtain active particles from very immature apples. As will be seen from Figs. 1 and 2, young apples are particularly rich in chlorogenic acid and tannins. We have, therefore, looked for some means of removing or rendering "harmless" phenolic compounds during the process of preparing mitochondria. Our early experiments showed that it was of little use merely to prevent oxidation of the interfering compounds.

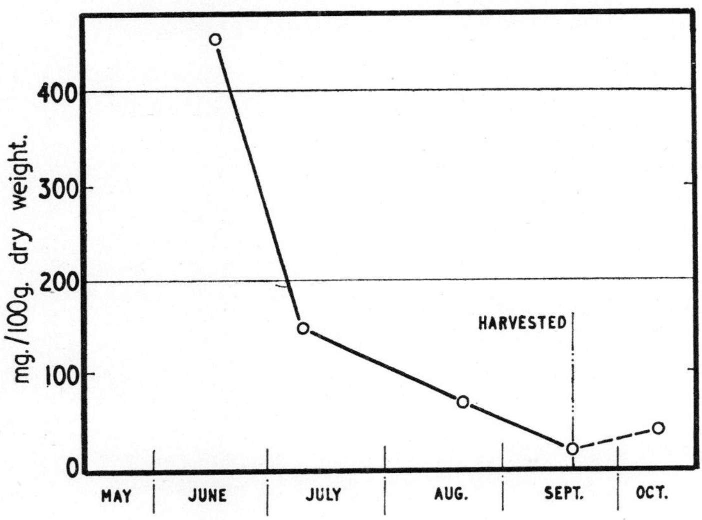

FIG. 1. The chlorogenic acid content of Cox's Orange Pippin apples during growth.

In work on the polyphenols of apple peel we had found that the polyamide powder "Ultramid" used by Grassman and others(5) for the separation of phenolic compounds would absorb most of the phenolic material in the peel. Unfortunately, when incorporated into the extraction procedure, "Ultramid" gave little improvement in the activity of the particles. Water soluble polymers reputed to cross link with tannins,(6) such as polyvinylpyrrolidone (PVP) looked promising, but our early use of this material was unsuccessful largely due to the presence of inhibitory substances in the PVP used. However, immediate success was achieved when we turned to a pharmaceutical grade of PVP. Our final technique, which made use of a specially designed mill(7) and incorporated vacuum infiltration of the tissue with buffer containing PVP before milling, gave high and reproducible activities even with young fruits. Figure 3 shows the effect of the use of PVP on the succinoxidase activity of mitochondria prepared from Cox's Orange Pippin apples picked in mid June.

It was found that the mitochondria contain some tannin even when PVP has been used during their preparation, and that the use of higher concentrations of PVP does not reduce the amount of this tannin. As might be

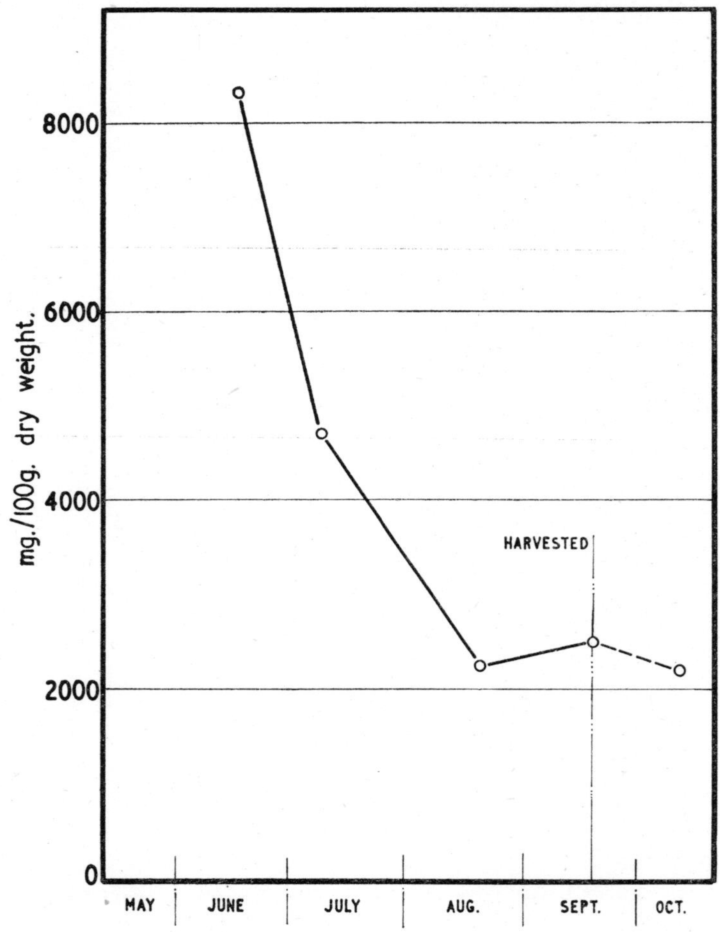

FIG. 2. The total tannin content of Cox's Orange Pippin apples during growth.

expected, the mitochondrial tannin content falls during the development of the fruit and the fall is paralleled by a fall in the total nitrogen content of the preparations. The magnitude of these changes will be seen from Fig. 4. It is clear that the general trend is the same both with and without PVP, but there is much more tannin in the preparations made without PVP, especially during the first 60–70 days from 1 June.

The succinoxidase activity of the preparations from whole fruits during the development on the tree are shown in Fig. 5. During the early days of June when tannin, and indeed nitrogen, are falling rapidly, the *activity per unit of tissue* also falls. But on a nitrogen basis (faint dotted line in Fig. 5) succinoxidase activity is rising and the rise continues up to maturity. This provides an explanation of the apparent paradox that activity per

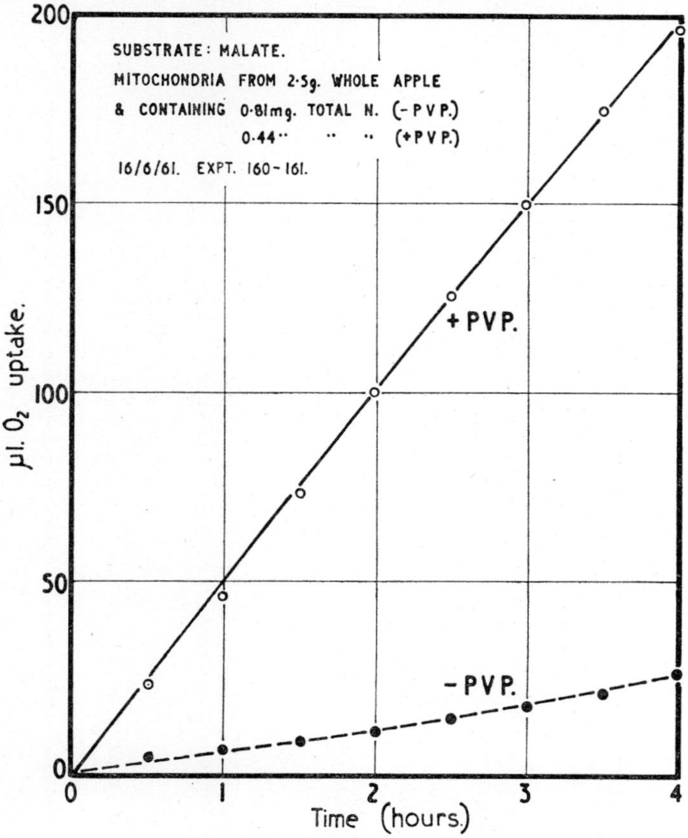

FIG. 3. The effect of PVP on the succinoxidase activity of apple fruit mitochondria.

unit of tissue is highest when the tannin content of the tissue and of the mitochondria prepared from it are also high. It is because the tannin brings down with the particles much inactive protein, especially in young fruits containing a great deal of tannin, even when PVP is used during the preparation.

In the young fruit, it appears to be the phenolic compounds which are removed by combination with PVP which exhibit a maximum inactivation of the mitochondrial activity, while the phenolics not removed by PVP are largely responsible for precipitating the inactive protein. This would explain why the activity does not appear to be greatly inhibited by the tannin which remains in the mitochondria after PVP treatment, and that

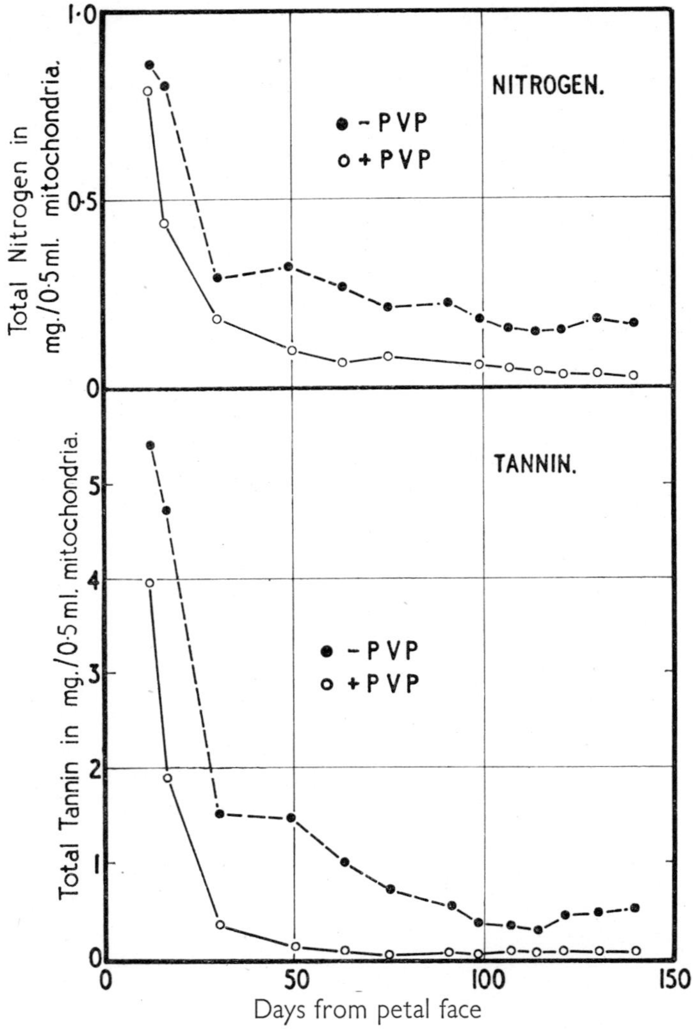

FIG. 4. Changes in the tannin content and nitrogen content of apple peel mitochondria prepared with and without the use of PVP.

when PVP is not used, whatever activity there is, is more susceptible to a high tannin content than to a high nitrogen content. It also means that in the very young fruits where the tannin content is very high, the tannins present are more of the type which cause direct inactivation rather than those which merely reduce the apparent activity (on a nitrogen basis) by precipitating inactive protein. Preliminary results suggest that in these young fruits there is a preponderance of simple leuco-anthocyanins, and that these compounds are indeed prevented from becoming incorporated into the mitochondrial preparations by the use of PVP.

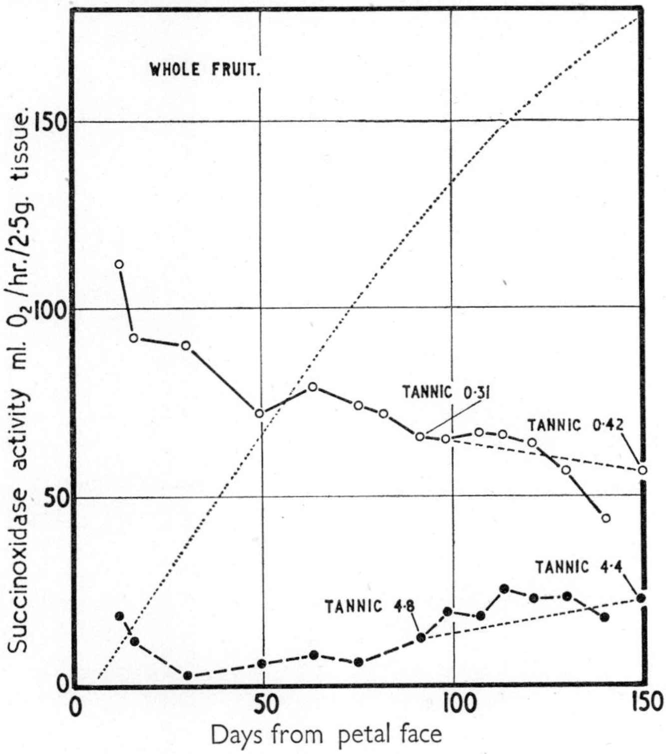

FIG. 5. The succinoxidase activity of apple mitochondria during the development of the fruit. Continuous line, prepared using PVP; interrupted line, prepared without PVP; dotted line, activity of mitochondria prepared using PVP, expressed per mg nitrogen. The tannic acid contents of stored samples, before and after storage (thinner interrupted line (are indicated in mg per 2·5 g original tissue.

It is clear from Fig. 5 that, in young apples, PVP has relatively more effect on activity than in more mature fruit. It will be seen later that when PVP is used with the peel of mature fruits, the nitrogen content of the preparations runs remarkably parallel with tannin content.

Lieberman[8] has found for mature American apples that for succinoxidase, the substrate reduction and the electron transport chain systems are most active if the mitochondria are prepared at pH values greater than 8. His activities were low in particles prepared at about pH 7, but he was not employing any "anti-tannin" ingredients apart from ascorbate to prevent oxidation. Lieberman's activities were all expressed in terms of the total nitrogen content of the preparation.

In the light of our results, it appeared possible that the high activities using alkaline extraction media obtained by Lieberman might be the result of a solubilization of any tannin–protein complex (or probably, rather, the prevention of its formation) so that non-mitochondrial protein would not appear in any quantity in the final particles. Their nitrogen content would

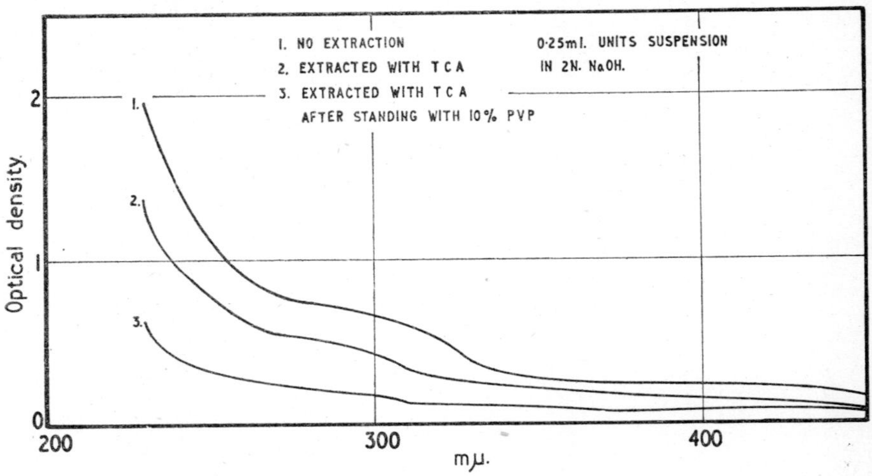

FIG. 6. Light absorption curves of alkaline solutions of mitochondrial preparations from apple peel (see text).

therefore be reduced and their activity *on a nitrogen basis* would be increased, although the absolute activity would not, in fact, be any higher. As we have shown elsewhere[1] this explanation appears to be a valid one.

We have compared the effect on their nitrogen content of extracting our mitochondria with trichloroacetic acid (TCA), perchloric acid, hot ethanol and lipid solvents but the results vary with the tannin content of the preparations. As yet we know practically nothing of the exact nature of the tannin–protein complex. That the complex can be dissociated by PVP even when it is already formed will be seen from Fig. 6, in which the light absorption is compared of a 2N-NaOH extract of: (1) peel mitochondria, (2) peel mitochondria after treatment with TCA, (3) peel mitochondria treated with TCA after standing for some time with PVP.

From the foregoing results it is quite clear that the phenolic compounds present in the fruit are largely responsible for reducing the activity of the isolated mitochondria, probably in two ways: (1) by direct inactivation and (2) by the precipitation of inactive protein which reduces the activity per unit of nitrogen present. The next stage is to examine further the action of PVP and to consider what phenolic compounds might be concerned in the inhibitory action. For this work it was desirable to use tissue which

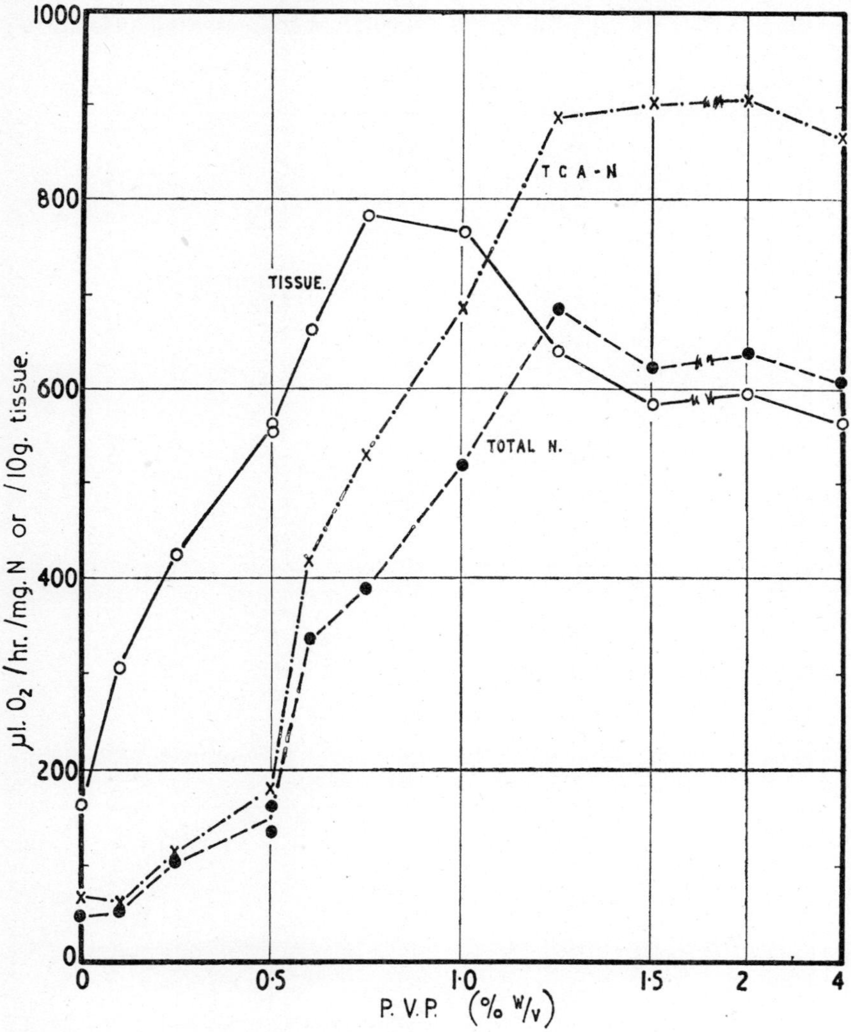

FIG. 7. Succinoxidase activity of mitochondria from apple peel prepared with various concentrations of PVP, expressed in terms of original tissue, per mg total N and per mg trichloracetic acid insol. N (TCA-N).

would yield mitochondria with a relatively constant potential activity but which, nevertheless, had a reasonably high tannin content. The peel of mature, post-climacteric apples was ideal for this purpose. It gave particles which had succinoxidase activity constant to less than 10 per cent and a tannin content, when prepared without the use of PVP, of about 1 mg compared with the 2–3 mg with mitochondria from whole, young apples. Such preparations were used in the results about to be described. As in the earlier work, the final pH of the extraction media was that usually employed for plant mitochondria, namely 7·0 to 7·5.

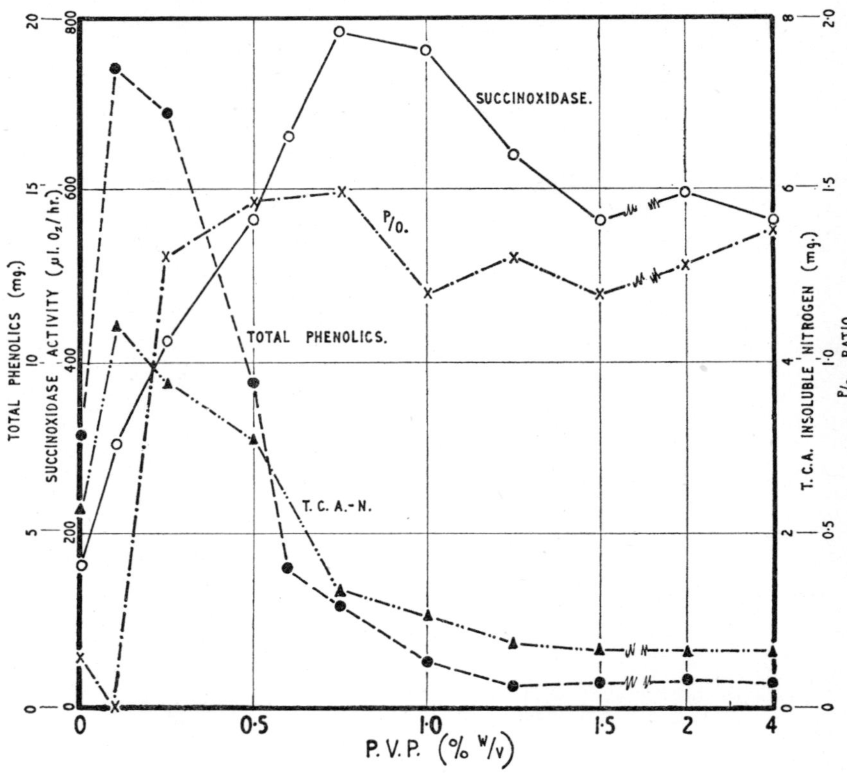

FIG. 8. Succinoxidase activity, TCA-N, tannin content and P/O ratios of mitochondria prepared from apple peel using various concentrations of PVP (see text). All the results are expressed per 10 g of original peel tissue.

Figure 7 shows the activity of mitochondria prepared in the presence of various concentrations of PVP. The interesting fact emerges that the optimal concentration of PVP appears to vary with the basis on which the succinoxidase activity is expressed and emphasizes the dilemma which arises when attempting to compare activities. Until it is clear what the nitrogen content of the various preparations signifies in terms of protein and enzyme, it is safer to compare activities in terms of the amount of tissue from which the particles have been isolated.

It is clear from Fig. 8, in which various properties of the mitochondria are plotted against the concentration of PVP used in their preparation, that the picture of inhibition of activity by tannin is not a simple one. In passing, it should be noted that in the region of the highest succinoxidase activity the ratio of phosphate uptake to oxygen uptake (the P/O ratio) is as high as 1·7; this is quite a high value for succinate oxidation by plant mitochondria. Minimum succinoxidase activity by no means coincides

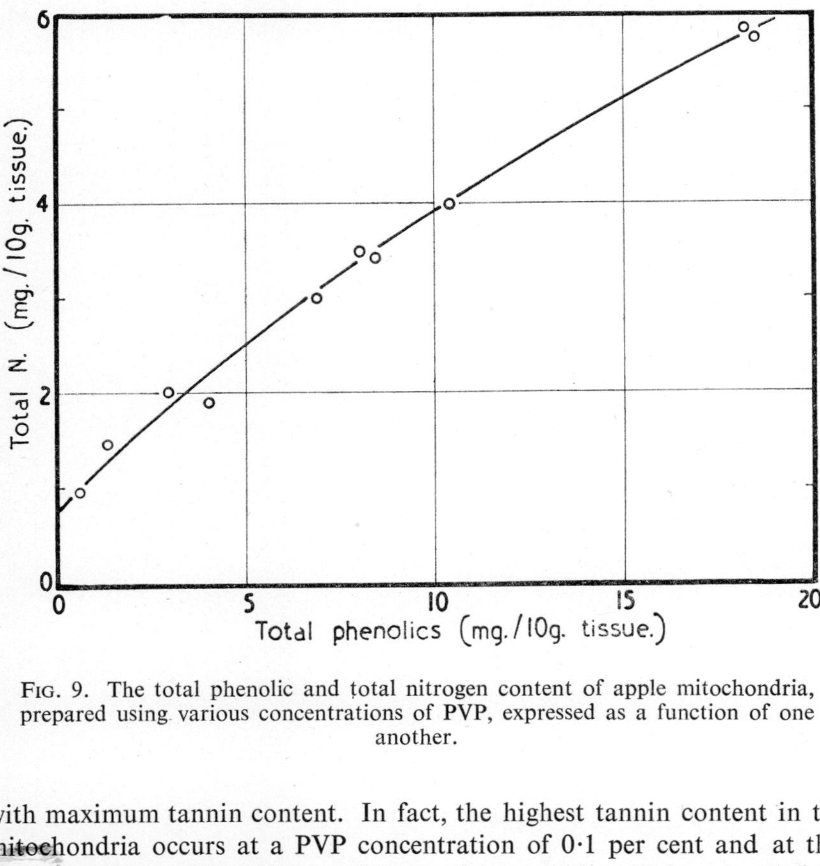

FIG. 9. The total phenolic and total nitrogen content of apple mitochondria, prepared using various concentrations of PVP, expressed as a function of one another.

with maximum tannin content. In fact, the highest tannin content in the mitochondria occurs at a PVP concentration of 0·1 per cent and at this concentration succinoxidase activity is rising rapidly. It is, nevertheless, true that the maximum succinoxidase activity is not reached until the tannin has fallen to a low value. The close parallelism between changes in tannin content and nitrogen content is maintained; this is quite clear from Fig. 9 in which nitrogen and tannin are plotted against one another over the range of PVP concentrations between 0 and 5 per cent.

Returning to Fig. 8, the optimal concentration of PVP for succinoxidase activity is 0·75 per cent and this appears to remain true for peel tissue throughout the storage life of the fruit used, and this concentration was used in all the subsequent work described here. Until more is known about the tannin–protein complexes and the effect of the various concentrations of PVP on their composition, the problem of the exact causes of the tannin inhibition and its reversal by PVP cannot be solved. At least we now have a satisfactory method for the study of the mitochondria and, probably, other enzyme systems of the apple fruit.

In the meantime some results have been obtained on the effect of a few phenolic compounds on the activity of peel mitochondria. The compounds tested are not necessarily those likely to be present in apples or indeed to have anything to do with mitochondrial activity *in vivo*. The work described is exploratory and the choice of phenolic compounds has been dictated to a large extent by their availability in sufficient quantity.

It became obvious from the start, and this is no new phenomenon,[(2)] that the mitochondria contained an active phenolase, and this has to be taken into account when considering the succinoxidase activity of the particles in terms of O_2-uptake (and in certain cases CO_2-output) in presence of oxidizable phenolic compounds. Obviously if it is the oxidized phenolics which are inhibitory, then a progressively increasing inhibition will be expected as the action proceeds. It is not yet clear whether the phenolase present on the particles is an integral part of them or whether it has merely become absorbed on their surfaces during the isolation procedure.

In the studies of the effect of PVP on the inhibition of succinoxidase by various phenolics which follow, the PVP (0·5–2·5 per cent) was added to the contents of the Warburg flasks in which the assay of succinoxidase activity was performed.

The relative activities of the phenolase present in the mitochondrial preparations with a variety of substrates are shown in Fig. 10. It appears that catechol, even at a concentration as low as 0·0005 M, is oxidized so rapidly that the phenolase is completely inhibited by the products of oxidation before the O_2-uptake can be measured by the Warburg technique. The process is slowed down by substitution at the 4-position of the benzene ring, e.g. using 4-methyl catechol. Nordihydroguaiacol is another compound related to catechol and this is also oxidized smoothly by the phenolase.

From the results shown in Table 1 it will be seen that the complete inhibition of succinoxidase by catechol is slightly reversed by PVP. The phenolase does not contain a cresolase although *p*-cresol inhibits the activity of succinoxidase to a considerable extent. Pyrogallol is oxidized by the particles and causes considerable inhibition of succinoxidase. The results for a number of phenolic acids are interesting. With gallic acid, itself actively oxidized by the mitochondria, there is considerable inactivation of succinoxidase which is partially reversed by PVP. The course of the reactions are shown in Fig. 11. When calculating the extent of the inhibition of succinoxidase activity from the results shown in Fig. 11 (and similar figures which follow), it is assumed that, where no inhibition takes place, the O_2-uptake in the presence of both phenolic compound and succinate should be the same as the sum of the O_2-uptakes of the phenolase

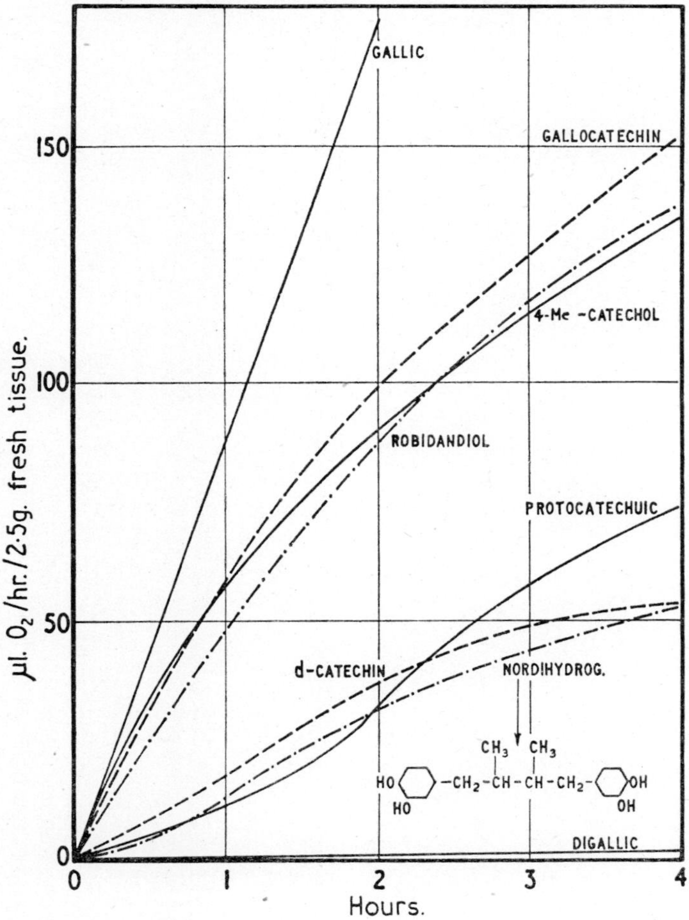

FIG. 10. The phenolase activity of apple peel mitochondria with various substrates all at a concentration of 0·005 M.

activity and the succinoxidase activity measured separately. Since gallic acid is so vigorously oxidized by the phenolase, it is possible that the inhibition of succinoxidase (which is partially reversed by PVP) is due largely to a competition for oxygen. It is somewhat difficult in these systems to assess the importance of a possible competition for oxygen. On the

TABLE 1. THE REACTIONS OF VARIOUS PHENOLS WITH THE PHENOLASE OF APPLE MITOCHONDRIA AND THEIR INHIBITIONS OF SUCCINOXIDASE

Compound	Conc. $\times 10^{-3}$ M	Hr	O_2-uptake (μl./2·5 g tissue)	CO_2-output (μl./2·5 g tissue)	Inhibition of succinoxidase (%) In absence of PVP O_2	In absence of PVP CO_2	In presence of PVP O_2	In presence of PVP CO_2
Catechol	1	1	0	0	98	100	90 (2·5%)	100
		4	0	0	100	100	96	91
		1	—	—	—	—	78 (5·0%)	85
		4	—	—	—	—	97	98
4-Methyl catechol	5	1	54	5·4	75	0	—	—
		4	22	2·4	89	100	—	—
Nordihydroguaiaretic acid	5	1	15	0	89	—	—	—
		4	2	0	99	—	—	—
p-Cresol	5	1	0	0	44	9	—	—
		4	0	0	53	67	—	—
4-Chloro-resorcinol	5	1	0	0	52	10	—	—
		4	0	0	86	79	—	—
Phloroglucinol	1	1	0	0	11	—	—	—
		4	0	0	15	—	—	—
Pyrogallol	2	1	98	28	36	19	—	—
		4	2	2	87	100	—	—

TABLE 2. THE REACTION OF VARIOUS PHENOLIC ACIDS WITH THE PHENOLASE OF APPLE MITOCHONDRIA, AND THEIR INHIBITION OF SUCCINOXIDASE

Compound	Conc. $\times 10^{-3}$ M	Hr	O_2-uptake (μl./2·5 g tissue)	CO_2-output (μl./2·5 g tissue)	Inhibition of Succinoxidase (%)			
					In absence of PVP		In presence of PVP	
					O_2	CO_2	O_2	CO_2
Protocatechuic acid	5	1	12	5·0	22	0	—	—
		4	16	2·6	72	100	—	—
Gallic acid	5	1	88	41	17	3	16 (0·5%)	0
		4	39	28	36	40	13	40
m-Digallic acid	5	1	22	8·0	71	−300	—	—
		4	0	0	92	75	—	—
Ellagic acid	5	1	0	0	21	—	—	—
		4	0	0	71	100	—	—

one hand, the cytochrome system operating in the Kreb's cycle has a much greater affinity for oxygen than has the copper–protein phenolase. On the other hand if the phenolase is absorbed on the outer surface of the mitochondrion it is much "nearer" to the ambient oxygen than is the cytochrome system which is within the mitochondrion assuming, of course, as has been suggested earlier, that the mitochondria are intact.

It is clear from Table 2 that both *m*-digallic and ellagic acids are virtually unattacked by the phenolase while both cause considerable inactivation of succinoxidase. There can be no question of competition for oxygen

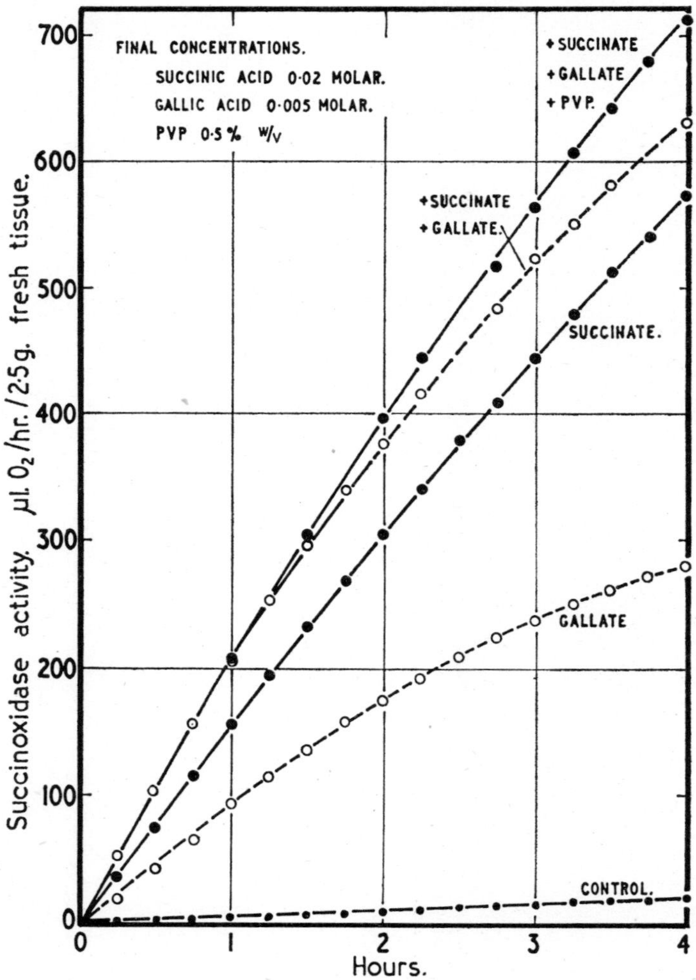

FIG. 11. The phenolase and succinoxidase activity of apple peel mitochondria and the effect of gallate on succinoxidase activity with and without the addition of PVP (see text).

here. The temporary increase in the CO_2-output of the Kreb's cycle system (a negative inhibition is an activation) on the addition of *m*-digallic acid is quite inexplicable.

It is interesting that chlorogenic acid at quite high concentrations (Fig. 12) while being itself oxidized, has only a slight inhibitory action on succinoxidase and the inhibition does not increase as oxidation of the

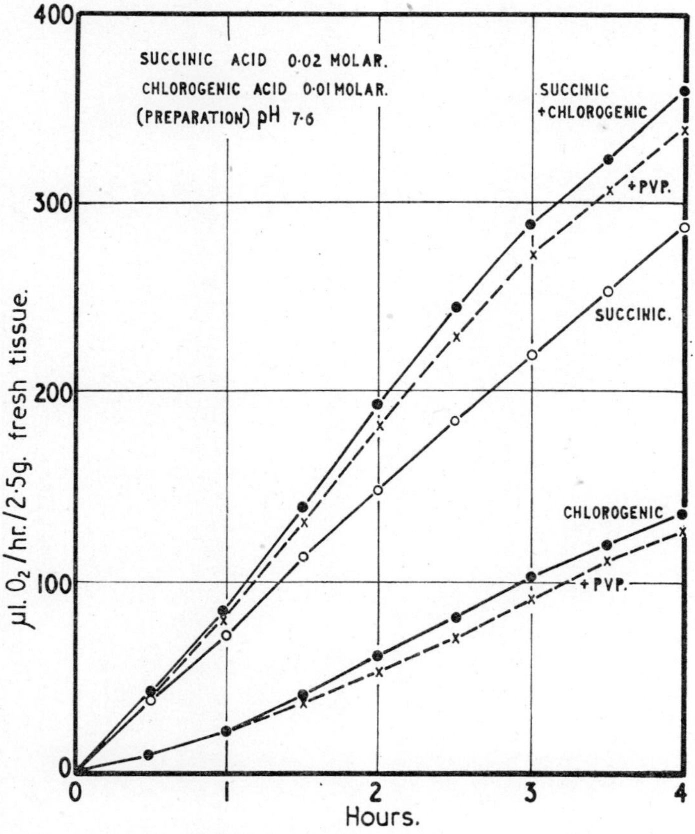

FIG. 12. The oxidation of chlorogenic acid and succinic acid, separately and together, by apple peel mitochondria with and without the addition of PVP (see text).

chlorogenic acid proceeds. This inactivation might possibly be merely a competition for oxygen; O_2-uptake with chlorogenic acid alone is proportional to substrate concentration between 0·01 M and 0·03 M. Oxidation of chlorogenic acid is hardly affected by PVP and therefore, presumably, it does not combine with PVP. It is interesting that caffeic and ferulic acids at a concentration of 0·005 M inhibit succinoxidase activity by about 40 per

TABLE 3. THE REACTION OF A COUMARIN AND OF ANTHOCYANINS WITH THE PHENOLASE OF APPLE MITOCHONDRIA, AND THEIR INHIBITION OF SUCCINOXIDASE

Compound*	Conc. $\times 10^{-3}$ M	Hr	O_2-uptake (μl./2·5 g tissue)	CO_2-output (μl./2/5 g tissue)	Inhibition of succinoxidase (%)			
					In absence of PVP		In presence of PVP	
					O_2	CO_2	O_2	CO_2
Aesculetin	5	1	0	0	32	38	33 (0·5%)	21
		4	0	0	55	28	56	20
Pelargonidin	9·7	1	—	—	70	76	—	—
		3	—	—	90	80	—	—
Pelargonin	5	1	—	—	40	0	—	—
		4	—	—	59	0	—	—
Cyanidin	5	1	—	—	17	20	14 (0·5%)	0
		4	—	—	51	11	33	0
Delphinidin	3	1	—	—	13	—	—	—
		4	—	—	10	—	—	—
Delphinin	3	1	—	—	21	−12	—	—
		4	—	—	13	−12	—	—

* With all these compounds excess Mn was present during assay.

TABLE 4. THE REACTION OF CATECHINS AND FLAVONOLS WITH THE PHENOLASE OF APPLE MITOCHONDRIA, AND THEIR INHIBITION OF SUCCINOXIDASE

Compound	Conc. $\times 10^{-3}$ M	Hr	O_2-uptake (μl/2·5 g tissue)	O_2-output (μl./2·5 g tissue)	Inhibition of succinoxidase (%)			
					In absence of PVP		In presence of PVP	
					O_2	CO_2	O_2	CO_2
d-Catechin	5	1	16	0	42	64	22 (0·5%)	−30
		4	3	0	69	100	11	35
	10	1	—	—	—	—	33	−30
		4	—	—	—	—	37	52
Gallocatechin	5	1	59	43	77	40	—	—
		4	27	17	86	75	—	—
	*5	1	41	6	70	68	47 (0·5%)	35
		4	11	0	87	90	50	25
Kaempferol	5	1	0	0	20	−30	—	—
Quercetin	5	1	5	0	19	−300	—	—
		4	7	0	70	100	—	—

* Excess Mn was present during assay.

cent; ferulic acid gradually inhibits the CO_2-output of the system completely.

The results for aesculin and several anthocyanins and anthocyanidins are shown in Table 3. As might be expected, none of these compounds is oxidized by the phenolase of the mitochondria. Nevertheless all of

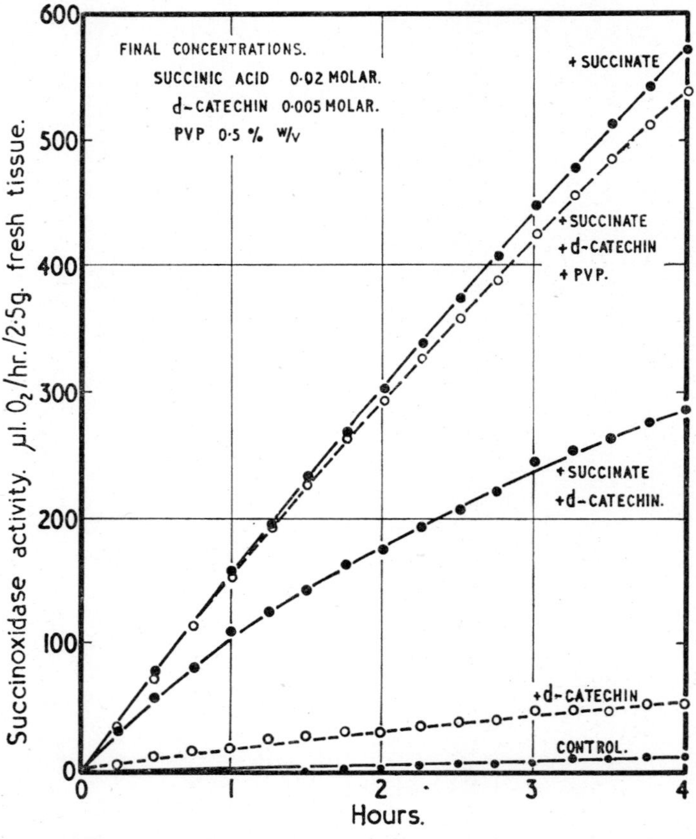

FIG. 13. The oxidation of *d*-catechin and succinic acid, separately and together, by apple peel mitochondria with and without the addition of PVP (see text).

them bring about some inhibition of both the O_2-uptake and the CO_2-output of the Kreb's cycle. The addition of PVP does bring about some alleviation of this inhibition but this is hardly significant.

The flavonols present a similar picture (Table 4), but the catechins give quite a different result inasmuch as they are oxidized by the phenolase. Data for *d*-catechin and gallocatechin are given in Fig. 13 and Table 4. It is clear that the flavonoid, *d*-catechin, is a powerful inhibitor of succinoxidase and the inhibition is almost completely removed by the addition of PVP. Both *d*-catechin and epicatechin are present in the apple fruit.[9] Gallocatechin is more rapidly oxidized by the phenolase of the mitochondria and also more severely inhibits the succinoxidase.

From the somewhat scanty data presented it appears that the catechin structure of the oxygen ring is more inhibitory to the succinoxidase than the corresponding anthocyanidin and flavonol structure and that the greater

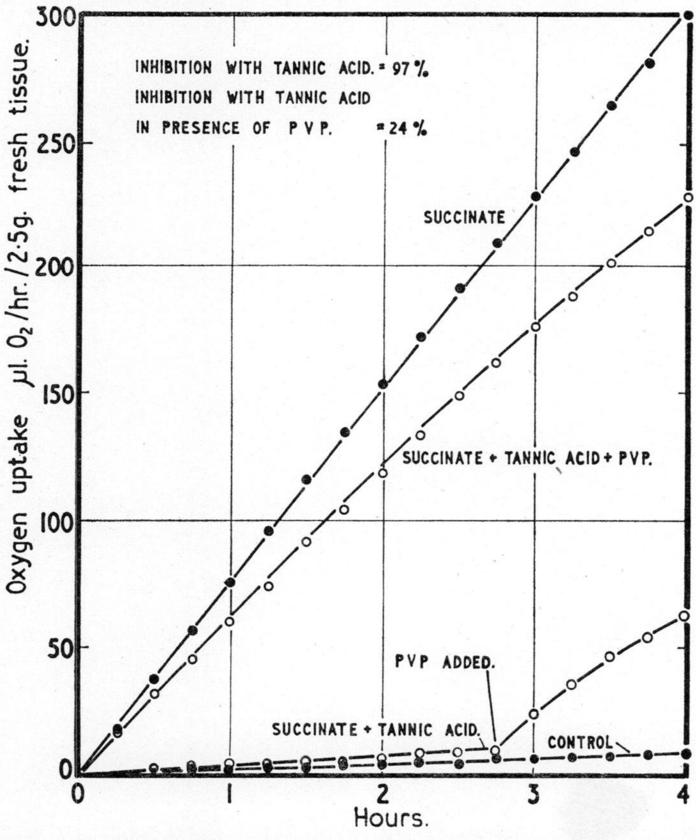

FIG. 14. The succinoxidase activity of apple peel mitochondria and the effect of tannic acid on this activity. The reversal of tannic acid inhibition by PVP. (Succinate, 0·02 M; tannic acid, 0·001 M; PVP, 2·5 per cent w/v.)

the degree of hydroxylation in the "B" ring the greater the inhibition. The evidence also suggests that the monomers of the phenolic acids and the flavonoid compounds are themselves inhibitory, i.e. that where inhibition does take place oxidative polymerization of the compound is not essential for inhibition.

When we come to the leuco-anthocyanins and the more polymerized phenolic compounds, we find an increasing inhibition with increasing

TABLE 5. THE REACTION OF LEUCO-ANTHOCYANINS AND TANNINS WITH THE PHENOLASE OF APPLE MITOCHONDRIA, AND THEIR INHIBITION OF SUCCINOXIDASE

Compound	Conc. $\times 10^{-3}$ M	Hr	O_2-uptake (μl/2·5 g tissue)	CO_2-output (μl/2.5 g tissue)	Inhibition of succinoxidase (%)			
					In absence of PVP		In presence of PVP	
					O_2	CO_2	O_2	CO_2
Robidandiol (Weinges)	5	1	5	0	19	−40	—	—
		4	20	3	46	60	—	—
Leuco-anthocyanin polymer (Forsyth[11])	0·25	1	—	—	88	41	—	—
		4	—	—	90	80	—	—
Tannic acid (Kahlbaum)	1	1	0	0	*98	96	29	32
		4	0	0	97	93	32	46
							(2.5%) 36	—
							†PEG 67	—
	0·1	1	—	—	21	60	0	—
		4	—	—	30	40	(1%) 13	—
(Merck)	0·5	1	—	—	97	88	—	—
		4	—	—	95	97	—	—
	0·5	1	—	—	—	—	19	20
		4	—	—	—	—	(0·5%) 9	10
Mimosa extract (Hathway)	0·1%	1	—	—	82	80	—	—
		4	—	—	78	65	—	—
	0·01%	1	—	—	9	25	—	—
		4	—	—	14	24	—	—
Apple extract (Flood)	4 mg solids	1	0	0	20	—	—	—
		4	0	0	11	—	—	—
	8 mg solids	1	3	0	30	39		
		3	6	23	35	34		

* Tannic acid inhibition not reversed by nicotine or caffeine
† PEG = 1 per cent polyethylene glycol.

degree of polymerization. This is clear from Table 5 in which the effect on succinoxidase activity is shown for a monomeric leuco-anthocyanidin (robidandiol),[10] a dimericcac ao leuco-cyaninidin,[11] tannic acid (Merck and Kahlbaum) and phenolic-containing extracts from both Mimosa bark and Cox's Orange Pippin apple.

The course of the inhibition of succinoxidase by Kahlbaum's "tannic acid" and its reversal by PVP are shown in Fig. 14. It is probable that the inhibition by tannic acid and other phenolic polymers allied to the leuco-anthocyanins is of the nature of a direct tanning action.[12]

TABLE 6. THE INHIBITION OF MALIC ENZYME BY VARIOUS PHENOLIC COMPOUNDS

Compound	Conc. $\times 10^{-3}$ M	Hr	% inhibition of malate decarboxylation
Chlorogenic acid	5	1	7
		2	− 6
		3	−29
		4	−63
d-Catechin	5	1	81
		2	88
		3	86
		4	63
Robidandiol	5	1	35
		2	35
		3	28
		4	0
Tannic acid	0·5	1	87
		2	91
		3	95
		4	90

Finally, it will be seen from Table 6 that tannic acid, *d*-catechin and robidandiol all have relatively the same effect on the malic enzyme of the mitochondria as they do on succinoxidase. Chlorogenic acid, however, has a progressively stimulating effect which needs further investigation.

We have carried out a few experiments on mitochondria prepared from tissues other than the apple, namely castor bean hypocotyls, etiolated pea seedlings, sprouting potato tubers and potato sprouts. Apart from the castor bean hypocotyls, these preparations were very low in total tannins and practically free from leuco-anthocyanins. The addition of PVP during the preparation of the mitochondria from the castor bean material reduced the tannin and nitrogen content of the particles and increased their succinoxidase activity to a considerable extent both on a nitrogen basis and per unit of original tissue. With the potato tubers and the etiolated pea seedlings the effect of PVP was marginal; it is proposed to investigate non-etiolated seedlings. With the potato sprouts, the material darkened rapidly during the preparation of the particles which had no Kreb's cycle activity. It would appear that PVP does not complex with the substrates of tyrosinase action.

Acknowledgements

We are indebted to Professor Weinges for a sample of robidandiol, to Dr. J. B. Harborne for various anthocyanidins and anthocyanins and to Mr. A. H. Williams for a sample of the Forsyth cacao leuco-cyanidin. Dr. D. E. Hathway kindly supplied the sample of Mimosa extract and the late Dr. E. A. H. Roberts a specimen of gallocatechin. Dr. A. E. Flood supplied the sample of apple extract.

REFERENCES

1. J. D. JONES and A. C. HULME, *Nature, Lond.* **191,** 370 (1961).
2. H. S. MASON, *Adv. Enzymol.* **16,** 105 (1955).
3. T. SWAIN and W. E. HILLIS, *J. Sci. Food Agric.* **10,** 63 (1959).
4. M. LIEBERMAN and J. B. BIALE, *Plant Physiol.* **31,** 420 (1956).
5. W. GRASSMAN, H. HÖRMANN and A. HARTL, *Makromol. Chem.* **21,** 37 (1956).
6. D. GUTTMAN and T. HIGUCHI, *J. Amer. Pharmaceut. Assoc.* (*Sci. Edn.*) **45,** 659 (1956).
7. J. D. JONES, in course of publication.
8. M. LIEBERMAN, *Plant. Physiol.* **35,** 796 (1960).
9. A. C. HULME, *Biochem. J.* **53,** 337 (1953).
10. K. FREUDENBERG and K. WEINGES, *Tetrahedron* **8,** 336 (1960).
11. W. G. C. FORSYTH and J. B. ROBERTS, *Chem. and Ind.* 755 (1958).
12. M. COLE, *Nature, Lond.* **181,** 1596 (1958).

CHAPTER 11

AROMATIC ACID-REDUCING SYSTEMS IN FUNGI

H. M. HURST

The Hartley Botanical Laboratories, University of Liverpool

THE TERM humic acid has been loosely applied to fractions obtained from soil by different extraction procedures and to products of different purity and, in view of the relatively few characteristics by which we recognize this compound and the unspecific procedures employed in its extraction, it almost certainly has different meanings to different research groups. Such information as we have on humic acid has recently been admirably reviewed by Kononova.[1] It is probably the most active component of soils where it occurs as a colloidal clay-humus complex with high base exchange capacity. Very little is known about the chemical nature of humic acid. Although a fundamental humic acid unit undoubtedly exists this is present in the soil as a complex with different cations, amino acids, peptides and other organic materials, and unspecific extraction procedures have led to the false inclusion of these entities as part of the basic unit. Significant advances in this field can only be achieved through a more detailed knowledge of the chemical structure of the basic unit and the establishment of procedures for its isolation and identification which will preclude the confusion which at present arises from a lack of criteria and parameters for characterization.

Savage and Stevenson[2] have justly pointed out that whereas information concerning the structure of complex biological materials such as polysaccharides, proteins and lignin has been obtained by hydrolytic and oxidative procedures which convert the material into units amenable to chemical study, no satisfactory procedure has yet been developed for the conversion in high yield of humic acid into its constituent parts. Such evidence as is available on the chemical nature of humic acid has led most workers to the tacit assumption that it is a heteropolycondensate of aromatic nuclei. There is, however, no single piece of evidence which conclusively proves this assumption and yields of aromatic compounds obtained by various degradative techniques are too low for reliable prediction of the structure of the basic humic acid unit.

Yields of crystaline material of 10 per cent which contain some nitrophenols and benzene carboxylic acids have been obtained by nitric acid oxidation of brown coal humic acids.[3] Unfortunately a close relationship between coal humic acid and soil humic acid has not been established and it would be wrong to extend this observation to cover humic acids obtained from soil. Morrison[4] has shown that several aromatic residues (vanillin, *p*-hydroxybenzaldehyde, syringaldehyde) can be isolated after alkaline nitrobenzene oxidation of humic acid. This observation does not establish humic acid as an aromatic compound, as the total yield of aromatic products was less than 1 per cent, which compares unfavourably with the yield of 20–40 per cent that may be obtained by alkaline nitrobenzene oxidation of lignin. Recent work by Greene and Steelink[5] has confirmed the presence of aromatic nuclei in humic acid. These workers used the milder CuO—NaOH oxidation described by Pearl and Beyer,[6] and showed that humic acid samples from three different podzols yielded similar oxidation products. Vanillin, *p*-hydroxybenzaldehyde, syringaldehyde, *m*- and *p*-hydroxybenzoic acids, vanillic acid and 3,5-dihydroxy benzoic acid were identified in the reaction mixture. Yields in this case were again low, about 1 per cent, but were comparable to those obtained by CuO–NaOH oxidation of lignin and these workers assume that aromatic compounds are significant structural elements of the humic acid molecule.

Evidence for the aromatic nature of humic acid has also been sought in its infrared spectrum. Farmer and Morrison[7] have shown that the infrared spectrum of humic acid from *Phragmites* peat has absorption bands clearly assignable to aromatic nuclei, but they indicate that this may be a particular case in which the parent plant material undergoes relatively little alteration during humification. Work at Liverpool has been predominantly centred on humic acid isolated from the B_1 horizon of a podzol beneath a plantation of *Pinus sylvestris*[8] and the infra-red spectrum of this product shows no absorption which can clearly be assigned to aromatic structures.

The recent isolation from soil samples of organisms which decompose humic acid[9] may provide a more fruitful line of enquiry in which biological degradation can be used as an alternative to conventional chemical degradation. One of these organisms *Polystictus versicolor* has been shown to decompose humic acid in liquid shake culture and on nutrient agar plates.[10] The rate of decomposition is slow and humic acid will not act as sole carbon source for this organism. In this respect decomposition of humic acid resembles decomposition of lignin, as lignin decomposition by *Polystictus versicolor* also requires an easily assimilable carbohydrate supplement.

Little is known of ring-fission of aromatic compounds by the higher fungi or if, indeed, such degradation occurs at all. The metabolism of various hydroxyl and methoxyl substituted benzoic acid derivatives has been investigated as a preliminary to a more detailed investigation of the pathway of humic acid decomposition.

Culture conditions included the pre-formed mat replacement technique and rapid shake culture technique. Chromatographic separations were carried out either on Whatman No. 4 paper using 1-butanol/ammonia (4:1, v/v), or on chromatostrips coated with Kieselgel G., using a chloroform (9:1, v/v) acetic acid solvent. Phenolic compounds were detected by spraying with diazotized sulphanilic acid followed by aqueous Na_2CO_3 solution.

P. versicolor does not metabolize aromatic acids by hydroxylation and subsequent oxidative cleavage of the aromatic nucleus and none of the intermediates usually associated with ring fission by micro-organisms could be detected. *P. versicolor* does, however, possess a marked ability to reduce aromatic acids and large quantities of the corresponding aldehydes or primary alcohols were identified by chromatography of growth solution extracts. This reduction of ring conjugated carboxylic acid groups has also been reported for *P. versicolor* by Farmer *et al.*(11) and Shimanzo and Nord.(12)

It is now known that this activity is shared by a number of other basidiomycetes and at least one ascomycete, and that there is an absolute correlation between the ability to decompose humic acid and to reduce the carboxyl groups of aromatic acids.(13) This correlation is clearly shown in Table 1. In 29 strains of fungi representing 20 species of basidiomycetes, 3 fungi imperfecti and 1 ascomycete there was an absolute correlation between the ability to decolourize humic acid and to reduce the carboxyl group of *m*-hydroxybenzoic acid.

Such knowledge as we have of the chemical structure of humic acid indicates that carboxylic acid groups are the most important surface reactive groups of the molecule, and these may well be reduced during decomposition. Evidence that this is the case was obtained by comparison of the infrared spectrum of humic acid in which carboxylic acid groups had been reduced by treatment with $LiAlH_4$, with humic acid which had undergone partial biological decomposition by *P. versicolor*.

A solution of humic acid in dry tetrahydrofuran (2 per cent, w/v; 250 ml) was refluxed for 5 hr with $LiAlH_4$ (1·0 g) and then left for 16 hr. Excess $LiAlH_4$ was decomposed with ethyl acetate and the reduced product now insoluble in tetrahydrofuran, was separated by centrifugation. Lithium and aluminium salts were removed by refluxing with 2N HCl (3 × 250 ml) and dialysing against running tap water for 2 days, and distilled water for 1 day, until there was no chloride reaction. After drying in a vacuum oven at 50°C, 4·7 g of the reduced product was recovered as a dark brown powder considerably lighter in colour than the starting material.

Lithium aluminium hydride reduces carboxylic acid groups as well as aldehydes and ketones and loss of the carboxylic acid group greatly alters the solubility of reduced humic acid. The reduced compound is insoluble in several organic solvents which will dissolve humic acid (Table 2). Its solubility in aqueous solution above pH 9·0 is probably due solely to dissociated phenolic hydroxyl groups.

TABLE 1. ACTION OF VARIOUS FUNGI ON HUMIC ACID AND *m*-HYDROXYBENZOIC ACID

Fungus	Humic acid decolourized	Aromatic carboxyl reduced
Polystictus versicolor (Linn.) Fr. Strain A	+	+
Strain B	+	+
Hypholoma fasciculare (Huds. ex Fr.) Kummer Strain A	+	+
Strain B	+	+
Strain C	+	+
Trametes suaveolens (Linn.) Fr. Strain A	+	+
Strain B	—	—
Trametes rubescens (A and S) Fr.	+	+
Polyporus brumalis Pers ex Fr.	+	+
Polyporus arcularius Batsch ex Fr.	+	+
Polyporus resinosus Schrad. ex Fr.	+	+
Polyporus schweinitzii Fr.	+	+
Ustulina zonata Lev.	+	+
Mycocalia denudata (Fr.) Palmer	—	—
Polyporus squamosus (Huds.) Fr.	—	—
Polyporus adustus Wild. ex Fr.	—	—
Polyporus betulinus Bull. ex Fr.	—	—
Polyporus dryophilus Berk.	—	—
Coniophora cerebella Pers.	—	—
Cyathus striatus (Huds.) Pers.	—	—
Collybia velutipes Curt. ex Fr.	—	—
Marasmius graminum (Libert) Berk.	—	—
Schizophyllum commune Fr.	—	—
Fomes pinicola (Sw.) Cooke	—	—
Poria ferrea (Pers.) Bourd. and Getz	—	—
Penicillium spinulosum Thom.	—	—
Aspergillus niger van Tiegh.	—	—
Penicillium janthinellum Biourgl	—	—
Boletus variegatus (Sw.) Fr.	—	—

Humic acid which had undergone partial decomposition was obtained by growing mature mats of *P. versicolor* on humic acid solution (0·12 per cent, w/v). After 10 days 74 per cent of the humic acid had been decomposed (determined by colorimetric assay of the growth solution[10]) and the residual was extracted with K_3PO_4 solution (10 per cent, w/v) and precipitated with HCl.

Humic acid absorbs strongly at 5·84 μ (Fig. 1) and this is assigned to carboxylic acid absorption rather than some other group. Farmer and Morrison[7] have shown that the intensity of this band is but little reduced when humic acids are treated with sodium borohydride, which reduces most aldehydes and ketones, but not carboxyl groups. Both $LiAlH_4$ reduced humic acid, and humic acid which has been partially decomposed

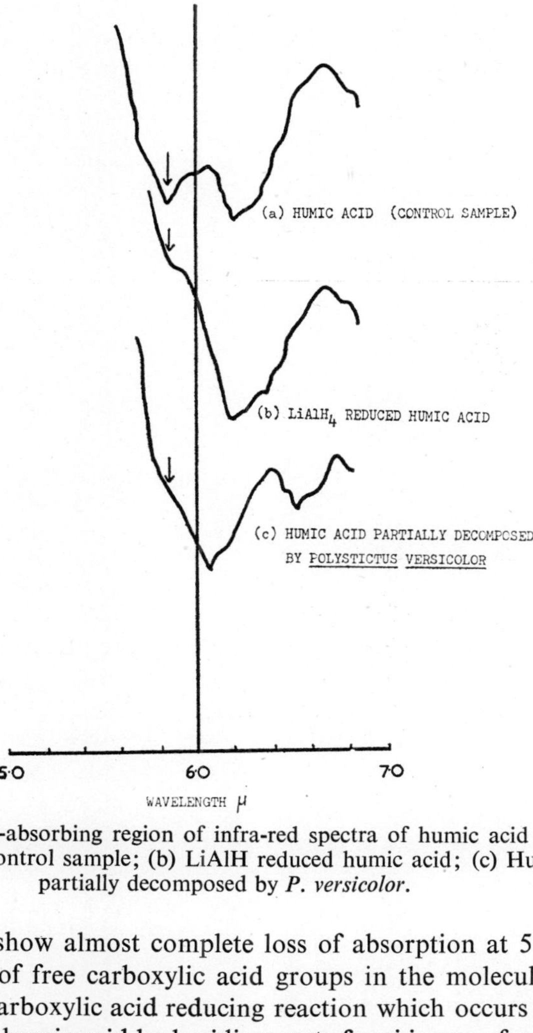

FIG. 1. Carbonyl-absorbing region of infra-red spectra of humic acid samples. (a) Humic acid control sample; (b) LiAlH reduced humic acid; (c) Humic acid partially decomposed by *P. versicolor*.

by *P. versicolor*, show almost complete loss of absorption at 5·84 μ indicating reduction of free carboxylic acid groups in the molecule (Fig. 1).

The aromatic carboxylic acid reducing reaction which occurs during the decomposition of humic acid by basidiomycete fungi is one of reduction of surface carboxylic acid groups of the molecule. This energy consuming reaction may be an early preparative step which facilitates further cleavage and separation of the constitutive units of the humic acid polymer.

P. versicolor will reduce a number of hydroxyl- and methoxyl-substituted aromatic acids including *o*-, *m*- and *p*-methoxybenzoic and β-napthoic acids;[11] 2,4- and 3,4-dimethoxybenzoic acids; and *m*- and *p*-hydroxybenzoic acids. Farmer *et al.*[11] suggest that the steric effect of ortho-substituents may interfere with the reduction of aromatic acids and cite 2,4-dimethoxybenzoic acid as inactive while the 3,4-isomer is readily reduced. The strain of *P. versicolor* used by us readily reduces 2,4-dimethoxybenzoic acid to 2,4-dimethoxybenzyl alcohol, which can be detected 6 hr after addition of the acid substrate to a mature culture. 2,4-dimethoxybenzyl alcohol gives a strong magenta coloured spot on

TABLE 2. CHANGE IN SOLUBILITY OF HUMIC ACID AFTER REDUCTION BY $LiAlH_4$ IN TETRAHYDROFURAN

Solvent	Humic acid	$LiAlH_4$ reduced humic acid
Tetrahydrofuran	+	—
Lactic acid (50%, w/v)	+	—
2-Methoxyethanol	+	—
0·2 M-sodium borate buffer pH 7·0	+	—
7·5	+	—
8·0	+	—
8·5	+	—
9·0	+	Slightly soluble
9·5	+	+

chromatostrips sprayed with diazotized sulphanilic acid and this is due to traces of nitrous acid in the spray reagent. Several dihydroxy- and dimethoxybenzyl alcohols were found to give strongly coloured spots with nitrous acid whereas the corresponding aldehydes and acids do not react. This reduction of 2,4-dimethoxybenzoic acid by *P. versicolor* and its inability to reduce the non-ortho-substituted compounds 3,5-dimethoxybenzoic acid and syringic acid suggests that some factor other than a steric effect of ortho-substituents is responsible for the absence of reduction in some compounds.

The reduction of aromatic acids by *P. versicolor* occurs in two steps and the corresponding aldehyde can always be detected in the growth solution at some stage prior to the appearance of the primary alcohol. The reduction is an energy consuming reaction, and, as in humic acid decomposition, reducing power is most evident in young, well aerated, liquid cultures with an easily assimilable carbohydrate. Under these conditions, chromatography of growth solution extracts shows clear evidence of the corresponding aldehyde, within 1 hr of addition of the parent acid substrate.

It has been established that a correlation exists between the two functions of aromatic acid reduction and humic acid decomposition, and the participation of a reducing step is further supported by the demonstration of the loss of carboxylic acid groups in humic acid partially decomposed by *P. versicolor*. Further evidence for the participation of an aromatic acid reducing reaction might be obtained by demonstrating the reduction of humic acid carboxyl groups by an isolated enzyme system which also reduces aromatic acids.

One of the two enzymes mediating this reduction, an aromatic aldehyde reductase, can easily be demonstrated in cell-free preparations of *P. versicolor*. Mycelium (15·0 g wet wt.) was ground in a pestle and mortar for 15 min with acid-washed sand (5·0 g) and 0·1 M potassium phosphate buffer (20·0 ml; pH 7·0). The slurry was filtered through nylon gauze and the crude extract centrifuged at 15,000 g for 15 min. The clear yellow supernatent was decanted off and used as the enzyme source. All operations were performed in a cold room at 2°. Reaction tubes contained enzyme solution (2·0 ml) and 0·08 M *m*-hydroxybenzaldehyde and 0·1 M potassium phosphate buffer (pH 7·0) in a total volume of 2·75 ml; the reaction mixtures were incubated for 1 hr at 25°. The reduced product, *m*-hydroxybenzyl alcohol, was detected by chromatography of ether extracts of the acidified reaction mixture.

A particulate fraction which probably corresponds to the mitochondrial fraction of higher plants, and which contains enzyme systems for the oxidation of Krebs cycle intermediates, including pyruvate, can be isolated from *P. versicolor*.[14] This fraction does not possess aromatic aldehyde reductase activity which is restricted to the 15,000 g supernatent fraction. Enzyme activity is lost after overnight dialysis against 0·1 M potassium phosphate buffer (pH 7·0) but can be restored by addition of 0·001 M $NADH_2$* or 0·001 M $NADPH_2$.* The enzyme is constitutive and activity is optimal between pH 7·0 and 7·5.

An enzyme which reduces benzaldehyde to benzyl alcohol, and which may be identical with that described here, has been obtained from cell-free preparations of yeast by Ose and Hironaka.[15] These workers were unable to determine if this enzyme was identical with yeast alcohol dehydrogenase.

Although the enzymatic reduction of aromatic aldehydes can easily be demonstrated in cell-free preparations of *P. versicolor*, we have been unable to demonstrate the corresponding reduction of aromatic acids. By analogy with known aliphatic carboxylic acid-reducing systems (e.g. the reduction of 3-phosphoglyceraldehyde during the reversal of the glycolytic pathway) this reaction is probably a complex one requiring preliminary activation

* $NADH_2$, reduced nicotinamide adenine dinucleotide; $NADPH_2$, reduced nicotinamide adenine dinucleotide phosphate.

of the substrate prior to reduction. Various combinations of the co-factors $NADPH_2$, $NADH_2$, coenzyme A and ATP did not, however, induce activity and the possibility that this enzyme is induced is now being investigated.

REFERENCES

1. M. M. KONONOVA, *Soil Organic Matter*, Pergamon Press, London (1961).
2. S. M. SAVAGE and F. J. STEVENSON, *Proc. Amer. Soil Sci. Soc.* **25,** 35 (1961).
3. W. FUCHS and W. STENGEL, *Annalen* **478,** 267 (1930).
4. R. I. MORRISON, *J. Soil. Sci.* **9,** 130 (1958).
5. G. GREENE and C. STEELINK, *J. Org. Chem.* **27,** 170 (1962).
6. J. A. PEARL and D. L. BEYER, *J. Amer. Chem. Soc.* **76,** 6106 (1954).
7. V. C. FARMER and R. J. MORRISON, *Sci. Proc. Roy. (Dublin Soc.* Ser. A, **1,** 85 (1960).
8. A. BURGES, *Sci. Proc. Roy. Dublin Soc.* Ser. A, **1,** 53 (1960).
9. A. BURGES and P. LATTER, *Nature, Lond.* **186,** 404 (1960).
10. P. LATTER and A. BURGES, *7th Int. Cong. Soil Sci.*, Madison, Vol. 2, 643 (1960).
11. V. C. FARMER, M. E. K. HENDERSON and J. D. RUSSEL, *Biochim. Biophys. Acta* **35,** 202 (1959).
12. K. SHIMANZO and F. F. NORD, *Arch. Biochem. Biophys.* **87,** 140 (1960).
13. H. M. HURST, A. BURGES and P. LATTER, *Phytocem.* **1,** 227 (1962).
14. D. BOULTER and H. M. HURST in *The Ecology of Soil Fungi*, p. 278, eds. D Parkinson and J. S. Wade, University Press, Liverpool (1960).
15. S. OSE and J. HIRONAKA, *Proc. Int. Symp. Enzyme Chem. Tokyo and Kyoto*, Vol. **2,** p. 457, Pergamon Press, London (1957).

CHAPTER 12

SOME NEW ENZYMIC REACTIONS OF PHENOLS

B. R. BROWN and SHEILA M. BOCKS

Dyson Perrins Laboratory, The University, Oxford

ONE OF our main interests during the past few years has been the determination of the structure of condensed, flavanoid vegetable tannins. It has become evident that existing chemical methods of degradation are incapable of yielding further information about tannin structure. We therefore decided to follow two lines of investigation:

(i) the development of new chemical techniques for degradation and

(ii) the discovery of enzymic methods for the stepwise breakdown of tannin molecules. The initial results of this latter approach are reported here.

Our initial substrate for this work was (+)–catechin (I), chosen because

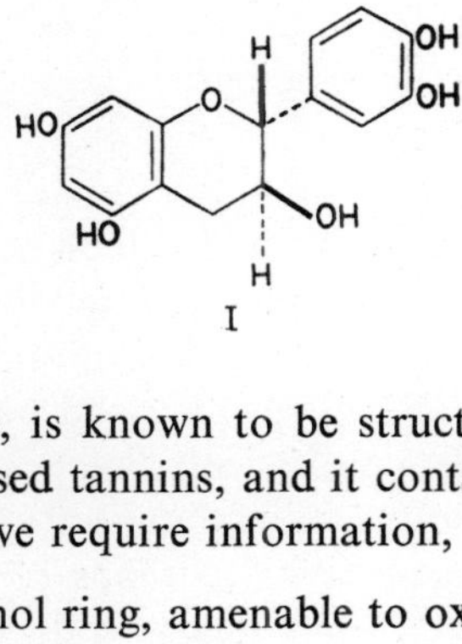

I

it is readily available, is known to be structurally close to the flavanoid monomers of condensed tannins, and it contains several sites for enzymic attack, about which we require information, viz.:

(a) the phloroglucinol ring, amenable to oxidative attack:[1]

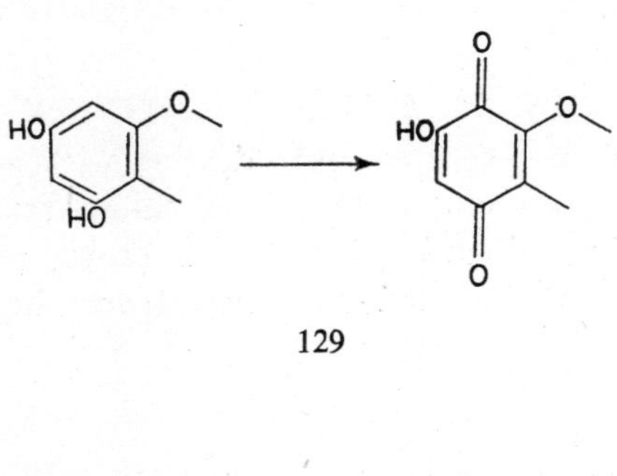

(b) the catechol ring, amenable to oxidative attack[2] and ring fission;

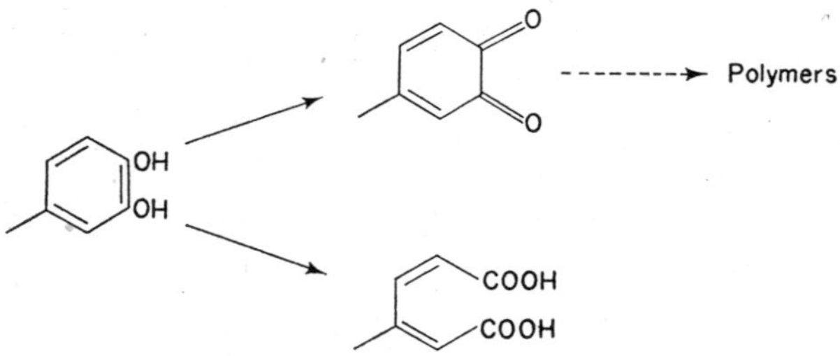

(c) the heterocyclic ether link, amenable to hydrolytic fission[3] or oxidative–hydrolytic fission:[4]

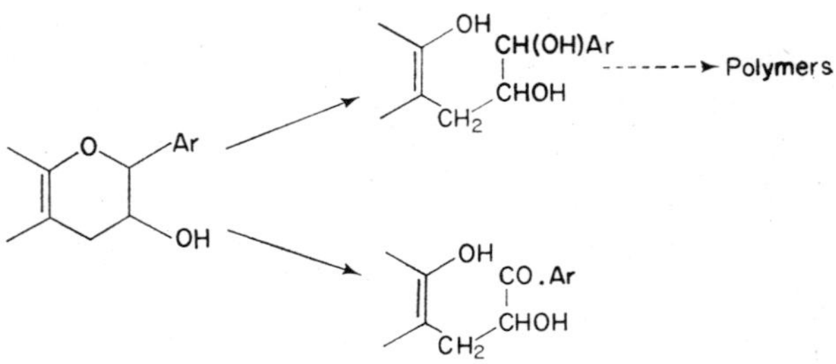

(d) the secondary alcoholic group, amenable to oxidation (a process not yet achieved chemically):

In our first experiment we used an unidentified soil fungus which grew on an agar–catechin medium and produced a yellow outer zone, indicative of the production of an extra-cellular enzyme. Following up this observation, we examined a number of known fungi and our results are collected in Table 1. It is clear that there are at least two types of reactivity: (i) the production of a yellow compound by the majority of the fungi; this we surmised was a phenol oxidase activity, perhaps of type (a) or (b) above and (ii) growth, but no coloured compounds produced; this we hoped was attack on the heterocyclic oxygen ring (type (c) above).

PHENOL OXIDASE ACTIVITY

We soon realized that here we were dealing with a phenomenon well described in the literature, viz. the Bavendamm test[5] which roughly differentiates white wood-rotting fungi from brown rots by the production, by the former, of coloured substances from phenols. Further, the extra-cellular enzyme from *Polyporus versicolor* responsible for this activity has been considerably purified by Fåhraeus and his co-workers.(6) However, as with other phenol oxidases, knowledge of the chemistry of the action of this enzyme was sadly lacking. This was the aspect which was of direct importance to our purpose.

TABLE 1. GROWTH OF FUNGI ON SOLID MEDIUM CONTAINING (+)-CATECHIN

Organism	Production of yellow compound	Growth
Polyporus versicolor	+++	Good
Polyporus sanguineus	+++	None
Polyporus hirsutus	+++	Poor
Polyporus abietinus	++	Poor
Stereum hirsutum	++	Poor
Marasmius scorodonius	+++	None
Trametes gibbosa	+++	None
Collybia butyracea	+	Poor
Hypholoma facicularis	+	None
Aspergillus niger	—	Poor
Penicillium solitum	—	Poor

At this point another line of research which we are pursuing made its impact: that of the synthesis of extended quinones, e.g. (II). We noticed

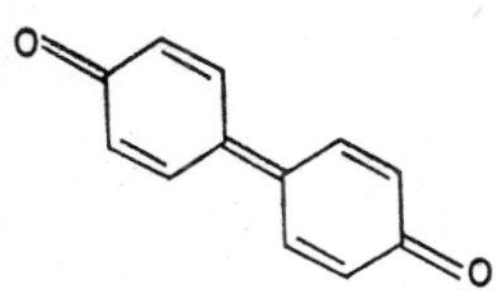

II

that Fåhraeus(7) has reported that growth of *Polyporus versicolor* on α-naphthol produces a purple colour and we knew from chemical oxidations of α-naphthol which we had carried out, that this purple substance was a quinonoid polymer which is very difficult to isolate or to investigate chemically. Consequently we decided to use a phenol substrate which was hindered in such a way that polymerization could not occur and from which an extended quinone (II) must result. (It is worth noting that many extended quinones which occur naturally are blocked in this way.) Our choice was therefore 2,6-disubstituted phenols (III), e.g. 2,6-dimethylphenol (III; R = Me) and 2,6-dimethoxyphenol (III; R = OMe). Growth of

Polyporus versicolor on these phenols produced the expected extended quinones (IV; R = Me or OMe) which were isolated and identified.

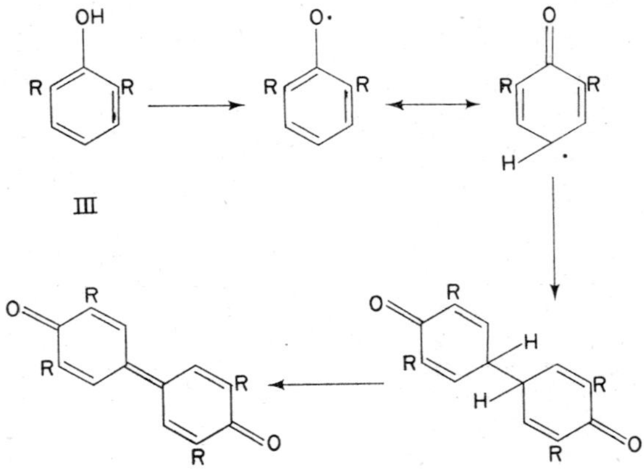

III

IV

The next step was to isolate the enzyme responsible for this reaction. For this we used the method described by Fåhraeus *et al.*[6] but with certain differences, the most important of which is our method of assay. For this we use 2,6-dimethoxyphenol at pH 4·0 and follow the reaction by the intensity of the ultraviolet band of the product, 3,5,3′,5′-tetramethoxydiphenoquinone (IV; R = OMe), at 468 mμ in chloroform. We chose this substrate because we have shown that its conversion to the diphenoquinone is quantitative and uncomplicated by side reactions or by further reactions of the product. This should be contrasted with methods which use catechol as substrate; catechol undergoes a complex reaction with this enzyme to yield at least five phenolic products of unknown structure. The dangers of using a complex reaction as a method of assay are illustrated by the following treatment. Suppose that a substrate (S), such as catechol, is chosen which undergoes further enzymic or non-enzymic reactions:

$$S + O_2 \xrightarrow[k]{\text{enzyme}} P$$

$$P + O_2 \xrightarrow[k_1]{\text{enzyme}} P_1$$

$$P + O_2 \xrightarrow[k_2]{\text{spontaneous}} P_2 \text{ etc.}$$

then, if k_1 and k_2 are comparable in magnitude with k and the reaction is followed by oxygen uptake or by the overall colour density of the products (P, P_1, P_2, etc.), the true activity of the enzyme (measured by k) is not being estimated. However, if a substrate, such as 2,6-dimethoxyphenol, is chosen which cannot undergo further reactions:

$$S + O_2 \xrightarrow[k]{\text{enzyme}} P \text{ (Final product)}$$

the above criticisms do not apply since the concentration of the initial product (P) is being measured directly.

A further advantage of 2,6-dimethoxyphenol as assay-substrate is that it should not be affected by tyrosinase[8] or by laccase (meaning by this term laccase from the lacquer tree which is said not to attack monophenols[9]).

The cell-free enzyme solution is obtained by the growth of *Polyporus versicolor* on a 3 per cent aqueous malt medium at 22°C for approximately three days, which corresponds to maximum enzyme activity. The resulting enzyme solution, after removal of the mycelium, is dialysed against 0·001 M-acetate buffer of pH 5·0, the enzyme is salted out by the addition of ammonium sulphate to saturation, dissolved in water, and again dialysed. The resulting solution is then used.

The specificity of the enzyme has been tested on about twenty phenols so far and the results are shown in Table 2. It is seen that several types of reaction can be differentiated:

(i) No reaction. This allows a differentiation from tyrosinase which readily attacks tyrosine.

(ii) Inter- or intra-molecular coupling to yield coloured extended quinones.

(iii) Inter- or intra-molecular coupling to yield colourless products.

(iv) Polymerization to yield complex structures and often complex mixtures of coloured substances.

(v) Pyrogallol yields the benztropolone structure, purpurogallin, which is susceptible to further attack.

It is of interest and of importance to note that this enzyme is sharply differentiated from laccase from the lacquer tree which is said to have no action on *p*-cresol or on resorcinol.[9]

The pH optima of the enzyme for a number of substrates have been determined and two optima seem to be favoured, viz. pH 3·6–4·0 (2,6-dimethoxyphenol and pyrogallol) and pH 5·0–5·4 (*p*-cresol, 2,6-dimethylphenol, catechin, and 2-methyl-1-naphthol). Two typical pH optima curves are shown in Fig. 1.

TABLE 2. ACTION OF AN OXIDASE FROM *Polyporus versicolor* ON PHENOLS

Phenol	Product
2,4-Dinitrophenol	No colour change
Tyrosine	No change (paper chromatogram)
2,6-Dimethylphenol	3,5,3′5′-Tetramethyldiphenoquinone
2,6-Dimethoxyphenol	3,5,3′,5′-Tetramethoxydiphenoquinone
2-Methyl-α-naphthol	3,3′-Dimethyldinaphthoquinone
p-Cresol	Pummerer's ketone (50%) and alkali-soluble polymers
2,4-Dimethylphenol	White solid
p-Methoxyphenol	Cream coloured solid
Griseophenone A	Dehydrogriseofulvin
Phenol	Pinkish-brown solid
Catechol	Brown solution; 5 phenolic spots on paper chromatogram
Guaiacol	Bluish-purple solid
o-Cresol	Brown solid
2,3-Dimethylphenol	Brown solid
Protocatechuic acid	Red solution; 1 phenolic spot on paper chromatogram
Gallic acid	Brown solution; phenolic streak on paper chromatogram
Catechin: pH 5·0	Yellow solid (λ_{max} 390 mμ)
pH 7·0	Orange solid (λ_{max} 460 mμ)
Resorcinol	Pink solid
Pyrogallol	Purpurogallin (100% in 27 hr)
Purpurogallin	4 Phenolic spots on paper chromatogram

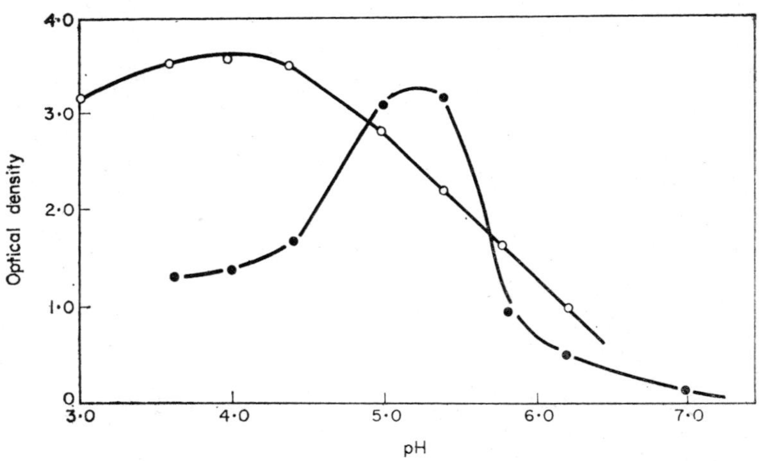

FIG. 1. Effect of pH on the activity of the *Polyporus versicolor* oxidase enzyme on (●) 2-methyl-1-naphthol and (○) 2,6-dimethoxyphenol.

The enzyme is a phenol oxidase which, by analogy with the known action of chemical oxidizing agents, e.g. potassium ferricyanide and manganese dioxide, is thought to function by the production of free radicals, which can couple from oxygen to *o*- or *p*-carbon atoms or from *o*- to *o*-,

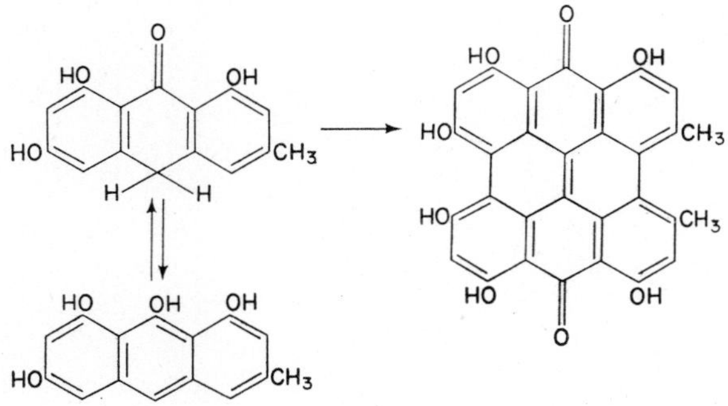

Hypericin

V

p- to *p*-, or *o*- to *p*-carbon atoms, followed by further changes dependent upon the chemical reactivity of the adducts so formed (see reaction schemes).

The production of extended quinones (IV) from phenols (III) is a reaction of importance in the biosynthesis of naturally occurring extended quinones, e.g. of hypericin (V).[10]

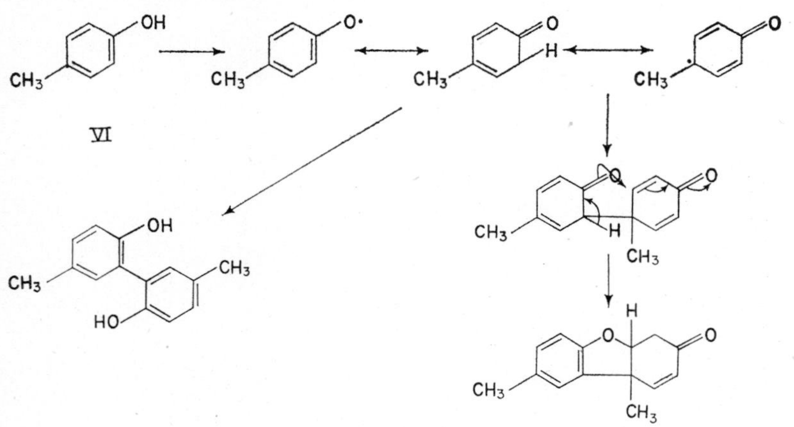

etc.

VII

Probably the most important reaction so far investigated is the oxidation of *p*-cresol (VI). The 50 per cent yield of Pummerer's ketone (VII) obtained should be compared with that of 13·4 per cent obtained by the use of hydrogen peroxide and peroxidase(11) and that of 20 per cent obtained by chemical oxidation with potassium ferricyanide.(12)

It is of interest that the Pummerer's ketone produced in this enzymic synthesis is optically inactive, whereas the phenol-coupled products synthesized by intact organisms (e.g. *Ramalina fraxinea* and *Usnea barbata* which produce usnic acid and *Penicillium patulum* and *Penicillium griseofulvum* which produce dehydrogriseofulvin) are optically active. This indicates either that the phenol-coupling enzyme in such organisms is more stereospecific than the one described here or that a second enzyme is utilized to cause stereospecific coupling of the initially formed "free" radicals.

However, the most important aspect of this reaction with *p*-creosol is that it has been regarded(13) as the prototype of a large number of phenol oxidations which are thought, in Nature, to lead to products of varied function. Thus we have shown that griseophenone A (VIII) is smoothly

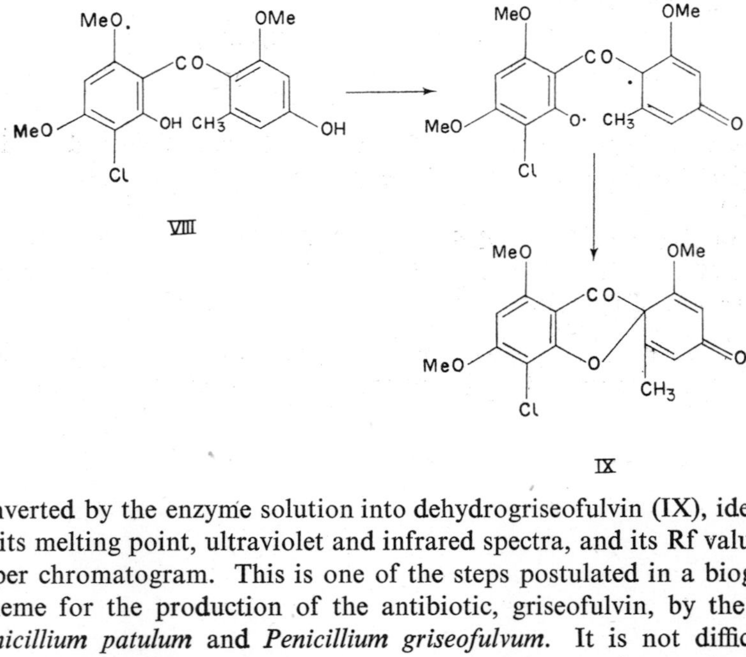

VIII

IX

converted by the enzyme solution into dehydrogriseofulvin (IX), identified by its melting point, ultraviolet and infrared spectra, and its Rf value on a paper chromatogram. This is one of the steps postulated in a biogenetic scheme for the production of the antibiotic, griseofulvin, by the fungi, *Penicillium patulum* and *Penicillium griseofulvum*. It is not difficult to envisage the application of this enzymic reaction to the production of other natural products and we have work in hand for the analogous synthesis of usnic acid, depsidones, morphine, etc.

It is paradoxical that having as our aim the isolation of enzyme systems capable of degrading complex polyphenols, we have characterized one which does the opposite, viz. one which produces polymeric products from simple phenols. Thus it seems likely that this enzyme would cause the production of artificial lignin from coniferyl alcohol.(14)

We think it possible that the white rot fungi elaborate this enzyme as a mechanism to remove toxic phenols from the growth medium, since these fungi habituate media which initially contain phenols and which have pH values on the acid side.

NON-OXIDATIVE ATTACK ON (+)-CATECHIN

In view of the above observations we realized that it was important to investigate those organisms which could grow on (+)-catechin but which showed no phenol oxidase activity, viz. *Aspergillus niger* and *Penicillium solitum* (see Table 1). When we examined the growth of these organisms on a liquid medium with (+)-catechin as the sole source of carbon, we found that, after several days, no phenolic compounds, apart from some polymer not corresponding to the whole of the (+)-catechin initially used, could be detected. Furthermore, no coloured compounds were produced, so that it appeared that phenol oxidase activity was not important here. Since we were looking for enzymes which would break the heterocyclic oxygen ring of catechin and eventually tannins, we next investigated the effect of stereochemistry at position 2 of the catechin nucleus on the utilization of catechins by these fungi. Epimerization of (+)-catechin (X)

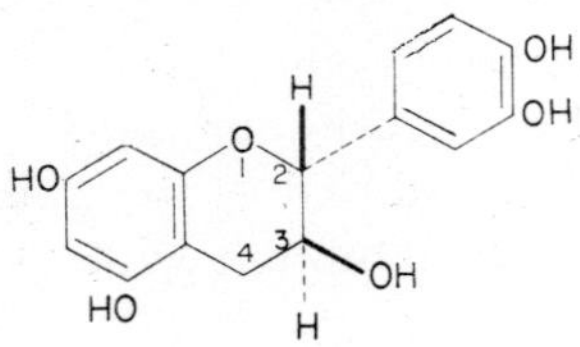

(+)-Catechin

X

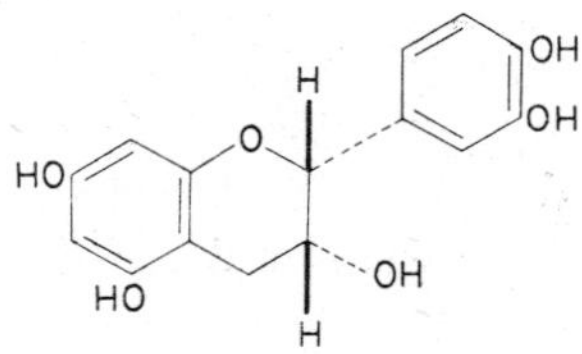

(−)-Epicatechin

XI

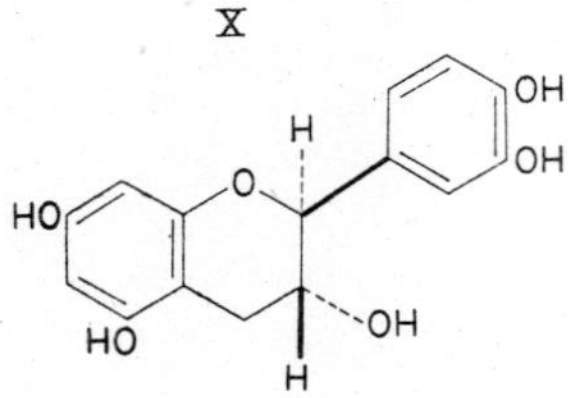

(−)-Catechin

XII

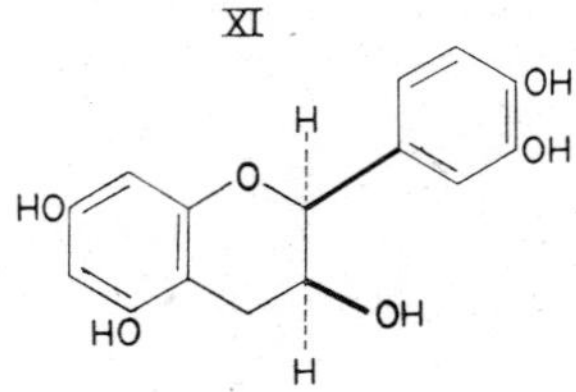

(+)-Epicatechin

XIII

produces a mixture of (+)-catechin and (+)-epicatechin (XIII), differing in their stereochemistry at position 2 and identifiable in presence of each other on paper chromatograms. Growth of *Aspergillus niger* or *Penicillium solitum* on this mixture resulted in complete disappearance of (+)-catechin (X) but persistence of (+)-epicatechin (XIII). Similarly, epimerization of

(−)-epicatechin (XI) gives a mixture of (−)-epicatechin and (−)-catechin (XII). These fungi remove (−)-epicatechin (XI) and leave (−)-catechin (XII). It is seen that the naturally occurring isomers (X and XI) which are attacked have the same stereochemistry at the 2-position and that the unnatural isomers (XII and XIII) which are not attacked, have the opposite stereochemistry at position 2. In the belief that this is controlling the degradation, we are at present engaged in isolating the enzyme responsible for the initial attack.

When *Penicillium solitum* was grown on epimerized (+)-catechin (10 g), we isolated, after methylation, a polymer (1·5 g), the tetramethyl ether of (+)-epicatechin (0·9 g), and a compound (0·35 g) which we believe to be

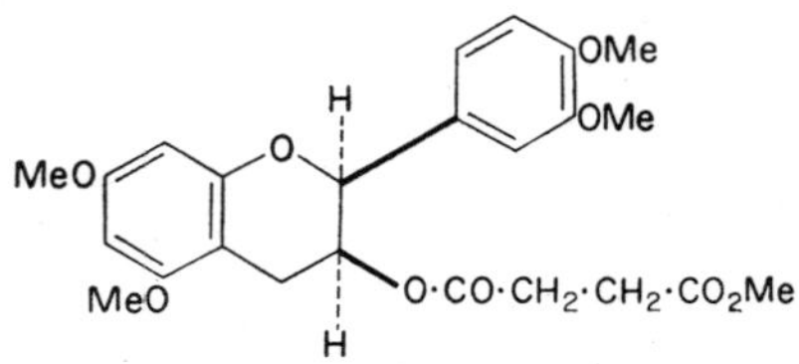

XIV

a succinic ester of (+)-epicatechin tetramethyl ether (XIV).[15] We are attempting to synthesize this compound to confirm our belief.

We would like to thank Messrs. G. Benfield and A. H. Todd for the work they have done and are doing in collaboration with us on these topics.

REFERENCES

1. ST. V. KOSTANECKI and J. TAMBOR, *Ber.* **35,** 1867 (1902).
2. D. E. HATHWAY and J. W. T. SEAKINS, *Nature* **176,** 218 (1955).
3. K. FREUDENBERG *et al., Annalen* **510,** 193 (1934); **590,** 140 (1954); *Chem. Ber.* **90,** 957 (1957); *Annalen* **612,** 78 (1958). B. R. BROWN *et al., J. Chem. Soc.* 3757 (1957); 4302 (1958); 3677 (1961).
4. *Cf.* J. AXELROD, *Biochem. J.* **63,** 634 (1956).
5. W. BAVENDAMM, *Ber. Deut. Bot. Ges.* **45,** 357 (1927).
6. G. MALMSTRÖM, G. FÅHRAEUS and R. MOSBACH, *Biochim. Biophys. Acta* **28,** 652 (1958).
7. G. FÅHRAEUS, *Kungl. Lantbr.-Högsk. Ann. Uppsala* **16,** 618 (1949).
8. M. L. CUSHING, *J. Amer. Chem. Soc.* **70,** 1184 (1948).
9. C. R. DAWSON and W. B. TARPLEY in *The Enzymes,* 1st Ed., Vol. 2, Part I, p. 484, eds. J. B. Sumner and K. Myrbäck, Academic Press, New York (1951).
10. H. BROCKMANN and H. EGGERS, *Angew. Chem.* 706 (1955); SHEILA M. BOCKS, B. R. BROWN and A. H. TODD, *Proc. Chem. Soc.* 117 (1962).
11. W. W. WESTERFIELD and C. LOWE, *J. Biol. Chem.* **145,** 463 (1942).
12. R. PUMMERER, D. MELAMED and H. PUTTFARCKEN, *Ber.* **55,** 3116 (1922).
13. D. H. R. BARTON and T. COHEN, *Festschrift A. Stoll,* p. 144, Birkhäuser, Basel (1957).
14. *Cf.* K. FREUDENBERG, *Chem. Ber.* **92,** 9C (1959).
15. M. M. BOKADIA, B. R. BROWN, D. COBERN, A. ROBERTS and G. A. SOMERFIELD, *J. Chem. Soc.* 1658 (1962).

INDEX